PENGUIN BOOKS

READING THE NUMBERS

Mary Blocksma has drawn on her ten
years of experience as a librarian to hunt
down the often elusive explanations
found in this book. She is the author of
sixteen children's books and lives in Holland, Michigan.

12345

READING THE NUMBERS

····

A Survival Guide to the Measurements, Numbers, and Sizes Encountered in Everyday Life

····

MARY BLOCKSMA

PENGUIN BOOKS

67890

PENGUIN BOOKS
Published by the Penguin Group
Viking Penguin Inc., 40 West 23rd Street,
New York, New York 10010, U.S.A.
Penguin Books Ltd, 27 Wrights Lane,
London W8 5TZ, England
Penguin Books Australia Ltd, Ringwood,
Victoria, Australia
Penguin Books Canada Ltd, 2801 John Street,
Markham, Ontario, Canada L3R 1B4
Penguin Books (N.Z.) Ltd, 182–190 Wairau Road,
Auckland 10, New Zealand

Penguin Books Ltd, Registered Offices:
Harmondsworth, Middlesex, England

First published in simultaneous hardcover and
paperback editions by Viking Penguin Inc. 1989
Published simultaneously in Canada

10 9 8 7 6 5 4 3 2 1

LIBRARY OF CONGRESS CATALOGING IN PUBLICATION DATA
Blocksma, Mary.
Reading the numbers : a survival guide to the measurements,
numbers, and sizes encountered in everyday life / Mary
Blocksma.
p. cm.
ISBN 0 14 01.0654 5
1. Weights and measures.   2. Weights and measures—United States.
I. Title.
QC88.B54   1989
530.8—dc19   88–23219

Printed in the United States of America
Set in ITC Garamond Book
Designed by Jane Treuhaft/Beth Tondreau Design

To Bruce,
With Love

CONTENTS

Acknowledgments xi
How Numbers Got Out of
    Hand xiii

Age, Extraordinary 3
Alcohol 5
Alcohol, Blood Levels 6
Annual Percentage Rate 7
Bar Codes 7
Barometric Pressure 12
Binoculars 15
Blood Pressure 16
Calcium 17
Calendar 19
Cans 22
Checks, Bank 23
Cholesterol 24
Cigarettes 26
Circles 30
Clothing Sizes 33
Comfort Index (Weather) 37

Commercial Items 40

Compass 41

Computers 43

Consumer Price Index (CPI) . . . 46

Copyright Page 49

Crash Test Rating Index
(CTRI) 51

Currency (Notes) 51

Distance, Nautical 54

Dow Jones Industrial Averages . 55

Dwellings (Size) 57

Earthquakes 58

Electricity 60

Engines (Horsepower) 63

Exponents 64

Fabric Care 66

Fabric Widths 67

Fertilizers 69

Financial Indexes 73

Firewood 74

Food (Energy Value) 78

Food Grading 81

Gas 82

Gasoline 85

Gold 88

Greenwich Mean Time 89

Gross National Product (GNP) . 89

Hats 92

Heart Rate (Pulse) 93

Highways 95

Humidity 99

Insulation 99

ISBN Numbers 105

Land Measures 106
Latitude and Longitude 107
Length (Common Short) 110
Light Beer 112
Light Bulbs 112
Lumber 114
The Metric System 116
Microwave Ovens 123
Military Time 123
Motorcycles 124
Nails 126
Oil (Engine) 128
Paper 129
Paper Clips 130
Pencils 131
pH 132
Pins 135
Plywood 136
Points, Mortgage 137
Postal Rates 138
Precious Stones 140
Prefixes, Astronomical 141
Prefixes for the Minuscule . . . 141
Prime Rate 143
Produce 144
Property (Legal Description) . . 145
Radio Waves 149
Rain 151
Roman Numerals 152
Rubber Bands 154
Sandpaper 155
Screws and Bolts 156
Ships 159

X CONTENTS

Shoes 161
Snow 162
Social Security Numbers 162
Socks 164
Sodium (Salt) 165
Soil, Garden 166
Sound 170
Staples 172
Steel Wool 173
Street Addresses 173
Sunscreen Lotion 174
Temperature 175
Tide Tables 177
Time 181
Time Units 183
Time Zones 184
Tires 188
Type 194
Universe (Distances) 196
Vision 198
Vitamins and Minerals 199
Volume 201
Week 203
Weight 203
Wind 206
ZIP Codes 208

Bibliography 211
Index 215

ACKNOWLEDGMENTS

I am deeply grateful to the many individuals, libraries, companies, and government agencies who responded generously and warmly to my many inquiries. I was especially astonished by the enthusiasm and support I received from persons at all levels of government—city, county, state, and federal—whose exceptionally patient and cheerful help have redefined the word "bureaucrat" for me.

Special thanks belong to the following:

AAA California; American Gas Association; American Heart Association; American Lung Association; American National Metric Council; American Paper Institute; Big O Tire Company; Building Division, City of Santa Rosa, Calif.; Bureau of Alcohol, Tobacco and Firearms; Bureau of Labor Statistics; California Department of Transportation; California Redwood Association; Chevron Research Company; Curtis and Turk, Healdsburg, Calif.; Federal Highway Administration; Federal Reserve System; Firestone Tire and Rubber Company; Hardin Mortgage Company, Santa Rosa, Calif.; Holland Public Library, Holland, Mich.; Hurty-Peck and Company; Industrial Fasteners Institute; National Association of Hosiery Manufacturers; University of California, Sonoma County Extension; National Academy of Sciences; National Bureau of Standards; National Fertilizer Development Center; Tennessee Valley Authority; National Highway Traffic Safety Administration; National Oceanic and Atmospheric Administration; National Office Products Association; National Weather Service; Office of Metric Programs, U.S. Department of Commerce; Pencil Makers Association;

Petrini's Market, Santa Rosa, Calif.; San Francisco Fashion Institute; Social Security Administration; Sonoma County Mapping Division; Sonoma County Public Library, Santa Rosa, Calif.; Tennessee Valley Authority; Tire Industry Safety Council; Uniform Code Council; U.S. Department of Energy; U.S. Geodetic Survey, Menlo Park, Calif.; U.S. Metric Association; U.S. Naval Observatory; U.S. Postal Service; U.S. Surgeon General's Office; and Wolverine Worldwide, Inc., Rockford, Mich.

Among the many individuals deserving special thanks are Valerie Antoine, executive director of the U.S. Metric Association; Carter Blocksma, Contour Designs, Grass Lake, Mich.; George Chrenka, Nu-Wool Company, Hudsonville, Mich.; Fred Dunnington, city planner of Middlebury, Vt.; John Gandy, Dorchester Industries, Shipyard Division, Dorchester, N.J.; Jack Herrod, General Pharmacy, Healdsburg, Calif.; Dr. William Klepczynski, Time Service Department, U.S. Naval Observatory; David McElroy, Oakridge National Laboratories, Tenn.; Chuck Northrup; Donna Preckshot; Dr. Fenwick Riley; Bob Ulrich, Modern Tire Dealer; and Charles Wilson of Industrial Fasteners Institute.

Finally, I wish to acknowledge those persons whose influence on this book was truly major: librarians Helen Hinterader and Jane Erla of the Healdsburg Public Library, who cheerfully chased down the answers to questions, month after month, bringing many new sources to my attention; my agent, Gina Maccoby; my wonderful Viking Penguin editors, Tracy Brown and Mindy Werner; and above all, Bruce W. Schadel, who tirelessly discussed, read, and critiqued this book and without whom I wouldn't have written it.

None of these persons, organizations, or agencies is responsible for any errors that may appear in this book; for those I take full responsibility.

M. B.

HOW NUMBERS GOT OUT OF HAND

Numbers seem to thrive in American life—it's impossible to avoid them. They come at us every day, from all directions, often at the most inconvenient of times. They find us on the highway and in the garden. They turn up at the gas station, bank, library, clothier's jeweler's, doctor's office, supermarket, fabric store, office supply store, pharmacy, and even at the beach. They fly off the financial pages, our gas bills, our personal checks, and practically every piece of packaging we touch.

Most of us go glassy-eyed before a selection of tires (P195/60R15?), sunscreens (6? 15?), light bulbs (Hours? Lumens?), or fertilizers (10-5-10 or 20-5-10?). And why not? No amount of American education—not a college degree, not even a Ph.D.—prepares us for these daily dilemmas. Numbers that we think we know, like the length of a yard and the time of day, are not always what they seem. Numbers on engine oil cans and those mileposts on the freeway have always been slightly ominous, at least to me. (Is there something here I should know?) The Consumer Price Index and the legal description on our property deeds have us on the run, not to mention those long, grim strings of numbers—checking account, bar code, ISBN, Social Security, and ZIP Code (plus an additional four numbers).

How on earth did numbers proliferate like this?

The first units of measure used on the earth were calculated quite naturally by every human body. An inch was the width of a thumb, a foot was the length of a foot, a pace was two marching steps, a mile a thousand paces. Liquid volume began with a mouthful, and that measure was doubled again and again for larger measures. It was a handy and a friendly system, and kept the universe within a person's grasp.

Effective as the body system was, it became apparent that some sort of standard had to be found so that people with different-sized extremities could communicate. The problem landed in the leaders' laps, and soon most of the world was measured by the hands and feet—and even the girth—of kings and queens. Unfortunately (or fortunately, in some cases), kings and queens were frequently replaced, and each time it happened, a country's system of measurements had to be revised.

In Europe frustrated traders and scientists began demanding measurements they could count on for longer than the royal life span, as well as more accuracy. Eventually standard platinum measuring instruments were made that were guarded like the royal jewels. Meanwhile, France devised the metric system, in which the meter was defined as one ten-millionth of the distance between the equator and the North Pole. Before the end of the nineteenth century, most of the world—except the United States and a few smaller countries—was metric.

It turned out, however, that the platinum meter became slightly distorted with time and temperature; worse, making copies of it was impossible. Scientists began searching for much larger, much smaller, and more accurate authorities. In 1960, the man-made meter was redefined in terms of the wavelength of light, then redefined again in 1983 using a laser. New measures like parsecs, microns, attograms and picoseconds began measuring an alarmingly expanding universe.

To cope with the numbers and information explosion, the computer was devised, and with it has come the binary number system which uses only zero and one. Most of us are spared direct exposure to numbers like 0001101, but we haven't been spared their impact. What used to be spelled out in plain English is now often coded in a strange language of numbers, from food at the market (bar codes), to books (ISBN numbers), ID cards, Social Security numbers, and checking account numbers. Numbers are becoming longer, more numerous, foisting themselves upon us without so much as a brief explanation.

Tired of being intimidated, and probably knowing less about numbers than you do, I began asking those truly simple questions one is usually too embarrassed to ask. The answers were astonishingly elusive. You would think, for example, that "How much alcohol is in this can of beer?" could be settled at a local package store. Often, however, such innocent inquiries required an astonishing amount of research, with calls to federal offices at the highest levels.

The result of this and hundreds of other such questions is a collection, arranged alphabetically, of the subjects that bothered me most. This book is not intended to be comprehensive or definitive—there are too many numbers to try to explain them all, and some are changing even as I write. The real purpose here is to take the "numb" out of numbers, hoping that you'll find the explanations useful, the histories amusing, and the oddities of many of these number systems vastly reassuring.

READING
THE
NUMBERS
. . . .

AGE, EXTRAORDINARY

The age of ancient objects, such as prehistoric bones and artifacts, and truly ancient objects, such as certain rocks and the earth itself, can often be calculated by analyzing the radioactive materials they contain. The calculations involve the radioactive elements' *half-life*, a term that may make your hair stand on end. Once you understand the idea, though, the dating process makes sense.

In 1900 Marie Curie discovered that certain rare elements (an element being a basic component of the universe, such as oxygen, hydrogen, carbon, sodium, etc.) become radioactive. At some point, a radioactive atom will begin to decay, giving off rays and particles, changing its basic structure, actually transforming into an entirely different element. This is truly startling behavior; we don't expect things to change their basic nature. Snakes do not change into birds, people don't become werewolves, and despite centuries of trying, lead (an element) cannot be turned into gold (another element). But a radioactive element continues transforming itself into other elements until it becomes a stable, nonradioactive element, and then it stops.

The timing of this transformation is also odd, because the individual atoms in a batch of radioactive materials are essentially ageless—they are no more apt to begin changing today than, in the case of uranium, for example, they were a million years ago. Nevertheless, it *can* be predicted that in 4½ billion years, half the atoms in any given amount of uranium-238 will have changed to thorium-234. Thus 4½ billion years is called uranium-238's *half-life*. This does not mean, however, that it will take only 4½ billion years to transform the half that's left. In 4½ billion more years, only *half* the remaining uranium will change, leaving ¼ of the original amount. It will take 4½ billion more years to transform half of that, leaving ⅛ of the original amount. And so on. It is because of this awkward-to-calculate tailing-off process that scientists use the easier-to-calculate half-life—or the time it takes to make the first split.

A half-life need not be billions of years. Thorium-234, for example, has a half-life of twenty-four days. In twenty-four days half of the uranium-238 atoms that decayed into thorium-234 atoms will have changed into protoactinium-234. The process continues until the uranium becomes lead, a stable, nonradioactive element. By measuring the amount of a radioactive material in a rock (or a very old artifact or bone), measuring the elements that this material turns into as it searches for stability, and then comparing these amounts, scientists can date ancient objects.

Different radioactive elements are useful for different ages. Uranium-238, with its extraordinarily long half-life, has been used to date rocks from 50 million to 4 billion years old—the earth itself has been dated at 5 billion years using this method. Potassium-40, with a half-life of 1.3 billion years, is often used to date younger rocks. Carbon-14, a radioactive element with a half-life of only 5,760 years (changing to carbon-12), is useful for dating prehistoric artifacts and bones—fossils of Neanderthal man have been dated at 50,000 years. Under 10,000 years, the carbon-14 method is accurate to within 100 years.

☞ The number following an element's name is its *atomic weight*. This number distinguishes an element from its *isotopes*—i.e., atoms with the same chemical properties but with a different weight. For example, carbon-14 is an isotope of the element carbon.

ALCOHOL

How do you tell how much alcohol is in what you're drinking? The percentage is clear enough on wine labels, but what is 100 proof brandy? And why is the alcohol content of some alcoholic beverages *not* on the label?

Wine and wine beverages are labeled with the percentage of alcohol content, but the actual percentage can vary 1.5 percent either way—a wine labeled 12 percent alcohol by volume may contain from 10.5 percent to 13.5 percent alcohol.

COMMON ALCOHOL CONTENTS

Wine coolers	3.5 to 7 percent
Table wines	7 to 14 percent
Dessert wines	14 to 21 percent
Distilled spirits	over 24 percent

Beer, by the federal government's decree, must contain at least .5 percent alcohol by volume to be considered beer. Most beers range between 4.4 percent and 5.5 percent alcohol, but not always. Inquiries into the amount of alcohol in beer opens a real Pandora's box. It seems that the federal government has the most liberal requirements about beer, but that any other government—state, county, even city or town—can tighten these up. Even states with no regulations at all may contain towns or counties that are regulated, or even dry. This means that the same brand of beer might have a 3.2 percent alcohol content if bought in Colorado, 4 percent in California, where anything over 4 percent must be sold as malt liquor, or 4.8 percent, the average for most beers, in states with higher ceilings.

Why isn't this information printed on the cans, like the alcohol content of wines? Because there's actually a law against labeling beer with the alcohol content. The reason given for this is that Congress feared that alcohol content labeling would lead to a sort of high octane race—beer companies would push beer by raising the alcohol content. How, then, do you know what strength beer you are getting? You don't. Often even the vendor doesn't know. (See also **Light Beer,** page 112.)

Liquor, such as whiskey, gin, brandy, etc., uses *proof degrees,* a confusing system used by the United States government. Fortunately, the math required to find the percentage of alcohol is simple: divide the proof by 2; e.g., 100 proof whiskey contains 50 percent alcohol. The highest proof possible, of course, is 200, or 100 percent alcohol. All alcoholic beverages are taxed on the basis of proof. This tax is included in the consumer's purchase price.

ALCOHOL, BLOOD LEVELS

Drinking and driving can be a fatal combination—a .10 percent Blood Alcohol Concentration (BAC) increases the odds of an automobile accident seven times—so the U.S. Department of Transportation has issued a chart to help drivers use good judgment. The chart not only shows how many drinks will make you too drunk to drive, it also shows that even half that level impairs reflex time and depth perception. Most state motor vehicle departments dispense charts and guidelines with every driver's license. If you've lost yours, use this one from the federal government.

BLOOD ALCOHOL CONCENTRATION CHART

Weight	Drinks (Two-Hour Period) 1½ ozs. 86 Proof Liquor or 12 ozs. Beer											
100	1	2	3	4	5	6	7	8	9	10	11	12
120	1	2	3	4	5	6	7	8	9	10	11	12
140	1	2	3	4	5	6	7	8	9	10	11	12
160	1	2	3	4	5	6	7	8	9	10	11	12
180	1	2	3	4	5	6	7	8	9	10	11	12
200	1	2	3	4	5	6	7	8	9	10	11	12
220	1	2	3	4	5	6	7	8	9	10	11	12
240	1	2	3	4	5	6	7	8	9	10	11	12

BE CAREFUL	DRIVING	DO NOT DRIVE
BAC to .05	IMPAIRED	.10 and Up
	.05–.09	

Source: U.S. Department of Transportation, National Highway Safety Administration.

To use the chart, select the line that applies to your weight. The white area is usually legal; the gray area is often illegal and certainly unsafe; the black area is illegal in most states and could be lethal. *Note:* These charts are guides and not legal evidence of actual blood alcohol concentration. Actual values can vary by body type, sex, health status, and other factors.

ANNUAL PERCENTAGE RATE

Every mortgage comes with two rates of interest—the nominal interest rate, or the one the loan is listed and advertised for, and the Annual Percentage Rate, known as an APR, which can be ½ percent or more higher. This sounds sneaky. Are you being bamboozled?

Here's what's going on: You decide on a mortgage with an interest rate of 10 percent, but there are expenses that may leave you fairly shaken if you are a first-time borrower. Points (prepaid interest; see **Points, Mortgage,** page 137) and/or the appraisal, the credit report, processing, and other fees (inclusions vary) add to the total cost of the loan. The APR is the interest rate calculated on the total of these plus the loan amount, while the nominal interest rate is calculated only on the amount of the loan you applied for.

Bamboozling has been outlawed in most states, which require lenders to inform you of the APR, particularly if it is appreciably *higher* than the nominal interest rate. If you aren't told, ask.

BAR CODES

What could be more inhuman than the grim black symbol stamped on nearly everything these days, most often encountered as the aptly named Universal Product Code (UPC)? Will it soon appear stamped on oranges, like the name Sunkist, or stuck on bananas like those little Chiquita stickers? Was it really necessary to make nearly every product on the market look incarcerated? Well, ask no more.

You are about to unlock those little bars. Bar Codes Can Be Fun.

One reason for the bar code fear factor is that the system hasn't been around long enough to feel really familiar. The UPC symbol wasn't introduced until 1973, when it was applied mainly to groceries, the industry that began the UPC movement. Now, of course, it's everywhere.

The Code. The UPC bar code is exactly that—a twelve-digit computer code that may be translated into Arabic numerals for humans. These twelve digits group into four numbers:

Source: *Uniform Code Council, Inc. Used with permission.*

0   The first digit, called the Number System character, identifies the product. A 0 is assigned to all nationally branded products except the following: a 2 signals random weight items, such as cheese or meat; a 3 means drug and certain health-related products; a 4 means products marked for price reduction by the retailer; a 5 signals cents-off coupons.

12345   The next five digits represent the manufacturer. This number is assigned by the Uniform Code Council (UCC) in Dayton, Ohio.

67890   The next five digits are assigned by the manufacturer to represent the product, and may include size, color, other important information.

5   The final digit is called the check digit. It signals the computer if one of the other digits is incorrect.

The price, by the way, is not part of most bar codes—that information is kept in the store's computer, which sends each price to the proper cash register as each bar code is scanned. Prices change, sales and specials are introduced, and prices vary from store to store, making it impractical to include the price in the bar code. Coupon clippers should know that the bar code not only automatically deducts the amount of the coupon from your bill but checks the grocery list to make sure you actually bought the product!

The Bars. The way the digits are turned into bars helps explain how the scanners "read" them. Each digit of the code is represented by two dark bars and two light "bars," or spaces, filling a space divided into seven equal parts.

SAMPLE "BAR" IN UPC SYMBOL

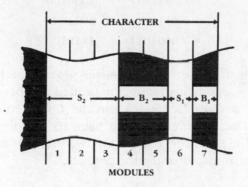

MODULES

S_1 and S_2 are the two groups of light bars; B_1 and B_2 are the two groups of dark bars.
Source: Uniform Code Council, Inc. Used with permission.

Because the computer can only read yes or no, each number is coded into a seven-digit yes/no code using a one (bar) for yes and zero (space) for no. It's easy to see how seven-digit code fills the spaces shown above.

There are two ways to write each number: A code applied to a digit in a manufacturer's number always contains an odd number of ones. This code is mirrored when applied to a product number digit—each one is turned into a zero, and each zero becomes a one, so that there is an even number of ones. (For example, a 5 is 0110001 in a manufacturer's number, but 1001110 in a product number.) This way, a manufacturer's number will not be confused with the product number. On page 10, find the code for 1 through 9, with those used for the manufacturer's code on the left and the mirrored version for product code on the right. Remember, each 0 is a space and each 1 is a bar.

Digit	Manuf. No.	Product No.
0	0001101	1110010
1	0011001	1100110
2	0010011	1101100
3	0111101	1000010
4	0100011	1011100
5	0110001	1001110
6	0101111	1010000
7	0111011	1000100
8	0110111	1001000
9	0001011	1110100

In a bar code, three sets of double bars separate each of the four number groups, extending below the printed numbers. The Number System Character (left side) and the Check Number (right side) are also represented by long bars just inside the outside double bars. The short bars encode the two five-digit numbers that represent the manufacturer and the product:

BREAKING THE BAR "CODE"

(NOT TO SCALE)

Right 5 Characters of Code

Left-Hand Guard Bars Pattern (101) — Left 5 Characters of Code — Tall Center Bar Pattern (01010) — Modulo Check Character — Right-Hand Guard Bar Pattern (101)

Number System Character

Left Light Margin Minimum 9 Modules Wide — Right Light Margin Minimum 9 Modules Wide

Number System Character — Modulo Check Character

1 2 3 4 5 6 7 8 9 0

Characters Per OCR-B Font

Source: Uniform Code Council, Inc. Used with permission.

The Scanner. In most supermarkets, the bar code is passed over a window, through which an electronic scanner using a laser translates the bars into time intervals, for it takes longer to pass over thick bars than thin ones. It measures spaces the same way. The time intervals are then translated into digits. Because the product codes are not the same as manufacturer codes, the laser can read the bar code in any direction. Laser light, by the way, doesn't break down into a rainbow of colors like ordinary light—it is only one color. Scanners use red lasers, which is why red ink cannot be used to print bar codes—the laser can't "see" red.

Check Number. Some people might like to know how the check number works—how can a one-digit number catch any mistakes in the preceding eleven? Information from the Uniform Code Council, Inc., says it works with this formula: Try it on our sample bar code number, always starting from the left. In this case, the odd positions begin with 0 and the even positions with 1:

0 12345 67890

1. Add the numbers in odd positions: $2 + 4 + 6 + 8 + 0 = 20$
2. Multiply total in Step 1 by 3: $3 \times 20 = 60$
3. Add the numbers in even positions: $1 + 3 + 5 + 7 + 9 = 25$
4. Add the Step 3 total to the Step 2 total: $60 + 25 = 85$
5. The number you need to add to reach a total that is divisible by 10 is your check number: $85 + 5 = 90$, so the check number is 5.

This really works—if numbers are transposed (a common mistake) or a wrong number is hit on a keyboard, the check number will be wrong, and the keyboard operator will know to check the figures. But because the check number is used mainly by computers, it's not always translated for the human eye into an Arabic numeral. You can still find the mistake, though, by using this formula.

BAROMETRIC PRESSURE

When the barometric pressure reads 30.1, what does this number
mean?—30.1 what? And why is the range—29 to 31 on many home
barometers—so small? You might assume that it means pounds of
pressure or something metric for which an American equivalent
was never found.

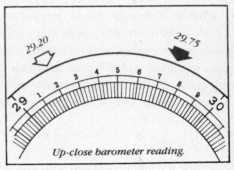

Up-close barometer reading.

This barometer reads 29.75, up from the last reading of 29.20.
Source: Howard Miller Clock Company, Zeeland, Mich.

In 1643 a student of Galileo named Torricelli took a tube about
34 inches long that was closed at one end, filled it with mercury,
turned it upside down, and put the open end in a mercury pool. No
matter when he did this experiment, the amount of mercury that
stayed in the tube ranged from about 29 to 31 inches, depending
on the air pressure.

Even though most of today's barometers are aneroid barometers,
using a dry partial vacuum to determine the air pressure, the reading represents the inches of mercury that would have remained in a
tube if mercury had been used instead. (Some barometers are also
marked with millibars. Millibar values represent the amount of
pressure exerted by the mercury column, with 1,013 millibars =
29.92 inches of mercury, or the mean pressure at sea level.)

How to Read the Barometer. The barometer uses decimal
inches—each number, or inch, is divided into tenths. The tenths

are divided into hundredths (although these are not marked on most home barometers). The barometric reading in the example is 29.42. Generally, a reading of over 30.2 is considered high and one below 29.7 low. Low air pressure brings inclement weather; high air pressure brings fair. To tell whether the air pressure is rising or falling (if the next reading is higher or lower than the last), line the stationary needle up with the barometer needle. Check later to see if the reading is higher or lower than the stationary needle.

If you know what direction the wind is coming from, you can take a stab at predicting the weather. As a rule, winds from the east quadrants and a falling barometer indicate foul weather. Winds shifting to the west quadrants indicate clearing and fair weather. How fast and how far the barometer falls can give you an idea how fast a storm is approaching and how intense it is. The following chart will hardly be enough on which to base accurate weather predictions—meteorologists still miss, even with the benefit of satellites and electronics. Still, it does give a feel for how those numbers, combined with a glance at your wind sock or weather vane, translate into reality:

WIND-BAROMETER TABLE

Wind Direction	Barometer, Normalized to Sea Level	Character of Weather This Indicates
SW to NW	30.10 to 30.20 and steady	Fair, with slight temp. changes for 1 to 2 days
	30.10 to 30.20 and rising rapidly	Fair, followed within 2 days by rain
	30.20 or above and stationary	Fair, with no decided temp. change
	30.20 or above and falling slowly	Slowly rising temp. and fair for 2 days
S to SE	30.10 to 30.20 and falling slowly	Rain within 24 hours
	30.10 to 30.20 and falling rapidly	Wind increasing in force, with rain within 12 to 24 hours
SE to NE	30.10 to 30.20 and falling slowly	Rain in 12 to 18 hours
	30.10 to 30.20 and falling rapidly	Increasing wind; rain within 12 hours

Wind Direction	Barometer, Normalized to Sea Level	Character of Weather This Indicates
E to NE	30.10 or above and falling slowly	In summer, with light winds, rain may not fall for several days. In winter, rain within 24 hours.
	30.10 or above and falling rapidly	In summer, rain probably within 12 to 24 hours. In winter, rain or snow with increasing winds possible.
SE to NE	30.00 or below and falling slowly	Rain will continue 1 to 2 days
	30.00 or below and falling rapidly	Rain, with high wind, followed within 36 hours by clearing, and in winter by colder temps.
S to SW	30.00 or below and rising slowly	Clearing within a few hours and fair for several days
S to E	29.80 or below and falling rapidly	Severe storm imminent, followed within 24 hours by clearing and in winter by colder temps.
E to N	29.80 or below and falling rapidly	Severe NE gale and heavy precipitation; in winter, heavy snow, followed by cold wave
Going to W	29.80 or below and rising rapidly	Clearing and colder

Source: "The Aneroid Barometer," U.S. Department of Commerce, Weather Bureau.

The Drastic Effects of Altitude on Barometers. If you moved to Denver from an area close to sea level, you might well assume your barometer was broken when you unpacked it—the needle wouldn't register anything. The atmospheric pressure is quite different at higher altitudes, and the readings of your barometer will be roughly 1 inch lower for every 1,000 feet higher you take it—even a trip of 100 feet will change the reading one-tenth of an inch. At 5,000 feet, the mean barometric pressure is 24.89 inches of mercury, a number so low that it's not even on your barometer.

How to fix this? A barometric pressure reading given in weather reports is usually "normalized" to what it would be at sea level. To fix your barometer, get a "normalized" report and reset your barometer needle to that reading. It won't be entirely accurate, since other factors affect barometric pressure readings, but it will be close.

☞ **Although barometers can be made from any liquid, those lighter than mercury require inconveniently lengthy tubes. A water barometer requires a 34-foot tube. The wine barometer once made by Blaise Pascal was 46 feet high.**

BINOCULARS

Binoculars come in such a strange variety of sizes—for instance, 6 × 15, 7 × 35, 8 × 30, or 10 × 50—that when you go to buy one, you can't imagine which one would be best. Is bigger better? Should you ask for the highest possible magnification? And what do those numbers mean, anyway?

Binoculars are sized by three basic numbers: the *power of magnification,* the *size of the "objective" lens,* and the *size of the field of view.* Magnification (the number before the "×") is the number of times the lens magnifies an object—anywhere from six to ten times. The size of the objective lens (the number after the "×") is the diameter of the large lens in millimeters—usually between 15 mm. and 80 mm. The field of view is usually measured in feet as seen from 1,000 yards away—typically 315 to 640 feet.

What you want from binoculars is a sharp, bright image, the result of a combination of magnification and lens size, and bigger is not, for most purposes, better. There are trade-offs: Higher magnification makes the image dimmer, so bigger objective lenses are needed to compensate. (The larger the lens, the more light it collects.) *But* bigger objective lenses make the binoculars bulkier and heavier. This is more than inconvenient—higher magnification makes it harder to hold a steady image, a problem made worse by heavy glasses. Thus 10× is about the practical limit.

Your intended binocular use will help you decide on a size. A reliably good size for birdwatching and all-purpose daytime use seems to be the middle-of-the-road 7 × 35, regular or wide-angle, while 7 × 50 works best for stargazing and other night use. Sizes larger than that—higher magnification and/or bigger lenses—work well for hunting. Smaller sizes are good for relatively close activities such as spectator sports, boating, and indoor use.

BLOOD PRESSURE

Until recently many doctors felt that your blood pressure was theirs to know and yours to worry about. Today, however, people are encouraged to find out their blood pressure. Blood pressure is always given in two numbers: *systolic* pressure over *diastolic* pressure: 120 over 80, for example. When people report high blood pressure, they often use the systolic—or high—number. Alarming as the number may sound, it's not usually the one used to measure "high blood pressure."

Systolic pressure (the high number) is the pressure of blood against the artery walls when the heart contracts. The diastolic pressure (low number) is the pressure of blood against the artery walls when the heart is at rest. The systolic pressure is always higher than the diastolic, because blood is pushed through the artery faster at the peak of the heart's pumping action. But it's the diastolic pressure that shows how much resistance has built up in the system, and it is this number that has usually concerned doctors most.

NORMAL BLOOD PRESSURE RANGES

Systolic pressure	120–160 millimeters
Diastolic pressure	50–90 millimeters

What do those numbers mean? While the *barometer* measures air pressure in inches of mercury, the *sphygmomanometer*—that pumped-up cuff that puts pressure on your upper arm—measures blood pressure in millimeters of mercury. Although pressure is now shown by a needle on a dial, it still represents the millimeters the mercury would show if a mercury sphygmomanometer had been used.

CALCIUM

If you're a woman, you're probably fairly annoyed by the confusing reports on calcium. You've been told that you're probably not getting enough of it, but understanding how much is sufficient is enough to make anyone fed up. Here's the problem: Not so long ago, the concern was over fiber. You were told that refined foods didn't contain enough fiber and that this could cause cancer of the colon. You learned that you should include fiber in your diet, found, in its most concentrated form, in bran. So you gave up eggs for breakfast (no fiber in eggs) and ate cereal.

That wasn't so bad. Then along came the bad news about cholesterol; what seemed like perfectly healthful food—meat and dairy products—was now full of a heart-threatening substance that clogged up your arteries. So you ate less cheese—you'd already cut out eggs during the fiber rush—skinned the chicken, and tried to live without hamburger.

And now it's calcium. But here's the rub: you've cut out all those dairy products, but lo and behold, dairy products are found to be the main source of calcium. Milk. Yogurt. Cottage cheese. It has also been reported that raw bran, recommended to increase the fiber in your diet, tends to absorb calcium. If you eat too much of it, say some experts, you counteract your efforts to build up your calcium levels.

How important is calcium? It's one of several factors that is associated with osteoporosis, a condition of weakening bones common in postmenopausal women. Osteoporosis, caused mainly by lack of estrogen, lack of exercise, and low levels of calcium, is most commonly seen in thin, light-boned, light-skinned women. In 1984, members of a National Institutes of Health conference on osteoporosis recommended that a woman increase her calcium intake from the average 450 to 550 mg. to 1,000 to 1,500 mg. per day well before menopause, unless she is prone to kidney stones.

Now, how is a health-conscious woman to get enough fiber, control cholesterol levels, and still maintain an appropriate level of calcium? It helps to keep things in perspective: According to many experts, calcium is the *least* important factor concerning osteoporosis, with estrogen and exercise far more influential. Here are a few suggestions:

1. To get enough fiber, eat fresh fruits and vegetables and yeast-leavened, whole-grained foods, such as whole wheat breads.

2. Drink skim milk—it is loaded with calcium (there are more than 300 mg. of calcium in an 8 oz. serving) and it contains none of the cholesterol found in 2 percent or whole milk. It also contains vitamin D, which helps your body absorb calcium.

3. Cheese is so high in calcium that small amounts on occasion shouldn't be too harmful, but eat foods other than cheeses that are high in calcium as well:

Foods High in Calcium	Approximate Mg. of Calcium
3.25 oz. sardines with bones	402
8 oz. skim milk	296
½ lb. raw turnip greens	558
1 cup bok choy	250
1 oz. Swiss cheese	251
1 7¾ oz. can salmon	339–570
1 oz. Cheddar cheese	213
2 oz. roasted almonds	213
1 cup cream of mushroom soup	190
½ cup collard greens	179
1 cup broccoli	136
4 oz. tofu	145
1 tbsp. blackstrap molasses	137
4½ oz. canned shrimp	147

(All vegetables listed are cooked.)

Source: U.S. Department of Agriculture.

Although there has been concern that some forms of calcium are more easily absorbed by the body than others, the latest studies (at this writing) have shown that you absorb about 20 to 26 percent of any calcium you take in, in whatever form.

4. Consider taking a calcium supplement. This can be somewhat confusing because calcium is sold for an astonishing array of prices and comes in several forms. The most common form is calcium carbonate and is perfectly acceptable unless you lack enough stomach acid, in which case, check with your doctor.

Calcium supplements may offer you anywhere from 100 mg. to 600 mg. per tablet, often for a similar price. Extra-strength Tums, containing 300 mg. of calcium per tablet, is an acceptable alternative if you prefer to chew your tablets. Not all antacids contribute to your calcium, however; some actually destroy some of the calcium in your system. Warning: A label proclaiming "750 mg. calcium carbonate" may translate into only 300 mg. of actual ("elemental") calcium—sometimes you have to read the fine print to find the amount of actual calcium.

5. Exercise! Moderate exercise really does help keep calcium from leaving your bones.

CALENDAR

There's an old joke that says the camel is convoluted because it was built by committee. Well, the camel is a keen piece of work compared to the calendar. The calendar was built by a whole string of committees, beginning with an ancient Babylonian committee and followed in later centuries by more committees appointed by pharaohs, emperors, caesars, and a pope. That date book of yours is actually an abysmal mess that survives only because you—and the rest of us—are blindly devoted to it. Unless scientists can sneak through a change behind our backs, as they did when they recently redefined the second (see **Time Units,** page 183), we won't let them alter our sacred sense of time.

For starters, our months, which are based on but don't really follow the moon's cycles, are of lengths so unpredictable that no Ph.D. has topped this nursery rhyme from the early 1600s:

> Thirty days hath September,
> April, June, and November;
> All the rest have thirty-one,
> Except February alone,
> Which has twenty-eight days clear
> And twenty-nine in each leap-year.
> —Anonymous

Now consider the weeks, which divide the months in a most erratic fashion, not even bothering to start each month with a fresh week. (See **Week,** page 203.) Only February divides evenly into weeks (except in leap years). Even the year doesn't divide evenly into fifty-two weeks—it's one day (two in leap years) too long. In addition, you need a new calendar every year, since the configuration is never the same two years running.

The reason for the confusion is that early on, committees began connecting time to the cycles of the earth, moon, and sun, which, though fairly predictable, are not really compatible:

A day   is based on the time it takes the earth to do a complete spin on its axis, although the time it takes can vary slightly.

A month   is based on the time it takes the moon to go around the earth, which is 29 days, 12 hours, 44 minutes, and 2.8 seconds.

A year   is based on the time it takes the earth to go around the sun, or about 365 days, 5 hours, 48 minutes, and 46 seconds.

Dividing solar time into dependable calendar units has long been a frustrating problem. An ancient Babylonian committee established a 354-day year, divided into 12 moon-cycles, each 29½ days long. This was eleven days short of the solar year, however, and their seasons began to drift. They fixed this by adding extra days to the calendar when it seemed to need some.

The Egyptians, still B.C., divided their year into twelve 30-day months, totaling 360 days, adding 5 days at the end to make 365. However, the year is actually more like 365¼ days long, so after centuries of use, their calendar was seriously off.

The Romans also had a lunar calendar, but they began with ten months—six had 31 days and four had 30 days, adding up to 304 days. The extra 61¼ days fell in winter when the priests must have assumed everyone was too depressed to notice, since they simply waited that long to announce the new year. Eventually the Romans added a couple more months, changing the number of days in each, until Julius Caesar managed to make twelve months add up to an exact 365¼-day year in 46 B.C. by adding a day—thus our leap year—every four years.

This would have worked if the year weren't actually eleven minutes shorter than that. By 1582 the seasons were so out of kilter that Pope Gregory XIII (see **Roman Numerals,** page 152) modified the calendar by refining the leap year—leap years would still be every fourth year, but years beginning a new century would not be leap years unless divisible by 400. (The year 2000, divisible by 400, will be a leap year. The year 1900 was not.)

Pope Gregory's calendar was only off by 26 seconds a year, which means it gets out of sync one day every 3,323 years. This has been an acceptable margin of error to most countries, which slowly began adopting the Gregorian calendar; Great Britain—and the American Colonies—adopted it in 1752. Then came Japan in 1873, China in 1912, with Russia balking until 1918. As you see, history's been hung up on it for more than 400 years.

Committees are still picking at the calendar even now. A new calendar has been suggested by the World Calendar Association that divides the year into four equal quarters of 13 weeks each, with a "World's Day" added at the end—an 8-day week at the end of December—to make 365 days (see next page). In leap years another World's Day would be added to the end of June as well. This calendar is so sensible that it probably will never catch on, but wouldn't it be wonderful to be able to use the same calendar every year?

WORLD CALENDAR ASSOCIATION CALENDAR

January

S	M	T	W	T	F	S
1	2	3	4	5	6	7
8	9	10	11	12	13	14
15	16	17	18	19	20	21
22	23	24	25	26	27	28
29	30	31				

February

S	M	T	W	T	F	S
				1	2	3	4
5	6	7	8	9	10	11
12	13	14	15	16	17	18
19	20	21	22	23	24	25
26	27	28	29	30		

March

S	M	T	W	T	F	S
					1	2
3	4	5	6	7	8	9
10	11	12	13	14	15	16
17	18	19	20	21	22	23
24	25	26	27	28	29	30

April

S	M	T	W	T	F	S
1	2	3	4	5	6	7
8	9	10	11	12	13	14
15	16	17	18	19	20	21
22	23	24	25	26	27	28
29	30	31				

May

S	M	T	W	T	F	S
			1	2	3	4
5	6	7	8	9	10	11
12	13	14	15	16	17	18
19	20	21	22	23	24	25
26	27	28	29	30		

June

S	M	T	W	T	F	S
					1	2
3	4	5	6	7	8	9
10	11	12	13	14	15	16
17	18	19	20	21	22	23
24	25	26	27	28	29	30 W

July

S	M	T	W	T	F	S
1	2	3	4	5	6	7
8	9	10	11	12	13	14
15	16	17	18	19	20	21
22	23	24	25	26	27	28
29	30	31				

August

S	M	T	W	T	F	S
			1	2	3	4
5	6	7	8	9	10	11
12	13	14	15	16	17	18
19	20	21	22	23	24	25
26	27	28	29	30		

September

S	M	T	W	T	F	S
					1	2
3	4	5	6	7	8	9
10	11	12	13	14	15	16
17	18	19	20	21	22	23
24	25	26	27	28	29	30

October

S	M	T	W	T	F	S
1	2	3	4	5	6	7
8	9	10	11	12	13	14
15	16	17	18	19	20	21
22	23	24	25	26	27	28
29	30	31				

November

S	M	T	W	T	F	S
			1	2	3	4
5	6	7	8	9	10	11
12	13	14	15	16	17	18
19	20	21	22	23	24	25
26	27	28	29	30		

December

S	M	T	W	T	F	S
					1	2
3	4	5	6	7	8	9
10	11	12	13	14	15	16
17	18	19	20	21	22	23
24	25	26	27	28	29	30 W

CANS

Occasionally you might come across an old recipe that calls for a No. 2 can of beans, but you're not sure how large a No. 2 can is. Today's cookbooks usually refer to a can size by ounces, but even that may change soon—the can is receiving a great deal of attention from the metric conversion people. Most cans today announce their amounts in both ounces and grams, but the day may not be far off when only grams appear on the label, with the size of the can influenced more by rounded metrics than cup/pint/quart amounts.

You're safe for the moment, however. Here are some ounce values for the more commonly called-for cans:

APPROXIMATE AMOUNTS IN CAN

No. 300	14 to 16 fluid oz.
No. 303	16 oz. to 17 fluid oz.
No. 1 tall	16 oz. fluid oz.
No. 2	1 lb. 4 oz. or 1 pint 2 fluid oz.
No. 2½	1 lb 13 oz. or 28 fluid oz.
No. 3	46 fluid oz.
No. 10	96 fluid oz.

CHECKS, BANK

When bank tellers ask for your checking account number, you know it's somewhere in that long string of strangely printed numbers on the bottom of your check. But which of those numbers do they want?

There is no straightforward answer. Often it's the last six digits. Sometimes it's more. Sometimes it's a group somewhere in the middle. To figure out your account number, write out the line of numbers from the bottom of your check in a large, friendly hand, substituting a pleasant space for the eye-bending squiggles.

Source: Federal Reserve Bank of New York, U.S. Department of the Treasury.

021302268   1234   07889   0687   0000010295

1. Except for the first number on top of the fraction, the numbers in the upper corners of the check are repeated in the bottom string. Find them and cross them out. Here is what they mean:

687	is your check number.
50	is the bank's American Banking Association number for the state or area in which your bank is located.
226	is your bank's identification number. (*Note:* It is the top of the fraction that needs to be listed when depositing checks.)
213	is the bank's Federal Reserve District, office and state or special collection arrangement.

2. 8   is the check digit (see **Bar Codes,** page 7). Cross it out.
3. 1234   represents your particular bank branch. Cross it out.
4. 0000010295 is the amount of money you wrote the check for, printed at the bottom of the check when it is returned with your statement, in this case, $102.95. Cross it out.
5. The block of numbers remaining is your account number (in this example, 07889).

CHOLESTEROL

Cholesterol has become a bad word in the American diet these days, and it doesn't seem fair—it is indispensable for the manufacture of sex hormones, not to mention brain and nervous system growth. Furthermore, although it's true that one kind of cholesterol clogs the inside of your arteries like layers of grease down the kitchen drainpipes, there's another type of cholesterol that does just the opposite—it actually helps clean the harmful kind out.

Still, there's reason to be concerned—bad cholesterol can cause atherosclerosis (hardening of the arteries), a major cause of heart disease. Most blood tests report the overall cholesterol present in your blood. The American Heart Association and most doctors now feel any count over 200 is undesirable—that's 200 mg. of cholesterol per deciliter of blood. If your level is between 200 and 239, you're considered to have "borderline-high blood cholesterol." Anything over 240 is considered high. Your risk of heart disease can be more accurately calculated, however, if another test is done which measures the clogging cholesterol, called LDL (low density lipoproteins), which may cause heart disease, and the declogging cholesterol, called HDL (high density lipoproteins), which tends to fight heart disease. The AHA recommends that everyone have this test done by the age of thirty (twenty, if there's a history of heart disease in your family), repeating it every three years.

It's risky to report the latest in cholesterol research—so many studies are being done, and the news does seem to change from year to year, if not from month to month. Studies have come out that show that stress can push blood cholesterol to instantly higher levels (Could needle-nervous people be showing higher serum cholesterol levels than stoics?); that women may make more "good cholesterol" (HDL) than men do (based on a study of female monkeys); that fish oils fight "bad" cholesterol; that fish oils don't fight bad cholesterol. One of the more disturbing studies showed that

CHOLESTEROL IN SOME FOODS
Recommended Daily Limit: 300 mg.

Food	Cholesterol Count
Chicken livers (3½ oz.)	631 mg.
Beef liver (3 oz.)	372 mg.
Egg, 1 medium	274 mg.
Chicken, skinless (3 oz.)	76–82 mg.
Ham, lean (3 oz.)	80 mg.
Hamburger, lean (¼ lb.)	77 mg.
Ice cream (1 cup)	51 mg.
Cheese, Cheddar (1 oz.)	30 mg.
Butter (1 tbsp.)	31 mg.
Milk, whole, 1 cup	33 mg.
Milk, skim, 1 cup	4 mg.

Source: American Heart Association.

cholesterol testing is not standard—that the same blood sent to many different laboratories came back with an astonishing range of cholesterol counts.

Despite these problems, it's only sensible to control your cholesterol. Most people can control their blood cholesterol by avoiding animal fats in their diets. American Heart Association studies show that the average American man consumes about 500 mg. of cholesterol every day, and the average woman 320 mg. The AHA recommends an intake of no more than 300 mg. daily. By cutting down on red meats, dairy products, and fast foods—some fast food restaurants fry almost everything in animal fat!—you may reduce your risk of heart disease.

CIGARETTES

Americans are becoming so sensible about smoking these days—quitting in droves, passing laws curbing public smoking, and, well, it's not so cool to smoke anymore—that the tobacco companies seem to be panicking. To fight possible disaster, lest this health trend continue, the tobacco industry has "lightened" up many of their products to promote the illusion that one can light up safely. They do this by adding righteous adjectives like "light" and "ultra light" to the names of their cigarettes. But if lights are really so much better than regulars, how come few tobacco companies put the details on the package?

There are three little numbers that most smokers would like to know—the amount of "tar," nicotine, and carbon monoxide contained in each cigarette for each brand and type. This information is really necessary if you want to know what you're smoking, since you can't go by the name—one company's ultra lights may contain the same amount of harmful substances as another company's lights or yet another's regulars.

On the average, the cigarettes at the "lightest" end of the scale contain about one-third the bad stuff of the filtered cigarettes at the strong end. But are cigarettes containing less tar, nicotine, and carbon monoxide less harmful? Possibly. But most smokers, trying to quit or cut back by smoking a "lighter" cigarette, simply smoke more of them. This must delight the tobacco companies, who have succeeded in getting many people to smoke more by smoking less. Furthermore, even cigarettes low in tar, nicotine, and carbon monoxide have been found by the Surgeon General to be disturbingly detrimental to health—one study shows nicotine to be as habit-forming as cocaine.

Obviously, the best thing for smokers to do is to quit smoking altogether. For smokers who won't quit, the Federal Trade Commission tested 207 varieties of domestic cigarettes for "tar," nicotine, and carbon monoxide. Because this information is not easy to find, a summary is included here (although all the figures may no longer apply, most will be close enough), despite its length.

MILLIGRAMS OF TAR, NICOTINE, AND CARBON MONOXIDE IN DOMESTIC CIGARETTES

Note: Measurements are given in milligrams (mg.) per cigarette; all cigarettes are filtered unless noted otherwise; carbon monoxide is shortened to CO; cigarettes are listed in increasing order of nicotine.

Less than 0.05 mg. tar, nicotine, and CO

Cambridge king size (hard pack)	Carlton 100 menthol (hard pack)
Carlton king size (hard pack)	Now king size (hard pack)

1–2 mg. tar, 0.1 mg. nicotine, less than 0.05 to 2 mg. CO

Benson & Hedges reg. size (hard pack)	Kool Ultra king size
Cambridge king size	Now king size, reg. and menthol
Carlton king size, reg. and menthol	Now 100, reg. and menthol
Carlton 100 (hard pack)	

3–4 mg. tar, 0.3 mg. nicotine, 2–4 mg. CO

Iceberg 100 menthol	Now 100
Kent III king size	Triumph king size, reg. and menthol

4–5 mg. tar, 0.4 mg. nicotine, 3–7 mg. CO

Benson & Hedges Ultra Light 100 (hard pack), reg. and menthol	Carlton 100, reg. and menthol
Cambridge 100	Doral II king size, reg. and menthol

4–5 mg. tar; 0.4 mg. nicotine, 3–7 mg. CO

Kent III 100

Kool Ultra 100 menthol

Merit Ultra Lights king size,
    reg. and menthol

Merit Ultra Lights 100, reg. and menthol

Salem Ultra king size menthol

Salem Ultra 100 menthol

Tareyton Lights king size

Triumph 100, reg. and menthol

True king size, reg. and menthol

Vantage Ultra Lights, reg. and menthol

Vantage Ultra Lights 100 menthol

Winston Ultra Lights king size

5–8 mg. tar, 0.5–0.6 mg. nicotine, 5–10 mg. CO

Bright king size menthol

Bright 100 menthol

Merit king size, reg. and menthol

Vantage Ultra Lights 100

Winston Ultra Lights 100

Belair 100 menthol

Carlton 120, reg. and menthol

More Lights 100 (hard pack)

More Lights 100 menthol (hard pack)

Pall Mall Extra Light king size

Parliament Lights king size (hard pack)

Parliament Lights king size

Tareyton Long Lights 100

True 100, reg. and menthol

Virginia Slims Lights 100,
    reg. and menthol (hard pack)

8–10 mg. tar, 0.7 mg. nicotine, 8–12 mg. CO

Belair king size menthol

Benson & Hedges Lights 100,
    reg. and menthol

Camel lights king size

Century Lights

Kent Golden Lights king size menthol

Kool Lights king size menthol

Kool Milds king size menthol

Kool Lights 100 menthol

L & M Lights king size

Marboro Lights king size

Marboro Lights 100

Merit 100, reg. and menthol

Newport Lights king size menthol

Pall Mall Lights 100

Raleigh Lights 100

Salem Lights king size menthol

Salem Slim Lights 100 menthol

Salem Lights 100 menthol

Vantage king size, reg. and menthol

Vantage 100, reg. and menthol

Viceroy Rich Lights king size

Winston Lights king size

8–12 mg. tar, 0.8 mg. nicotine, 6–13 mg. CO

Camel Lights 100

Kent Golden Lights king size

Kent Golden Lights 100, reg. and men-
    thol

Kool Milds 100 menthol

L & M Lights 100

Lucky Strike king size

Multifilter king size, reg. and menthol

Newport Lights 100 menthol

Old Gold Lights king size

Parliament Lights 100

Players king size, reg. and menthol

8–12 mg. tar, 0.8 mg. nicotine, 6–13 mg. CO

Raleigh Lights king size

Satin 100 menthol

Viceroy Rich Lights 100

Winston Lights 100

11–16 mg. tar, 0.9 mg. nicotine, 8–16 mg. CO

Alpine king size menthol

Century king size

Dorado king size

Eve Lights 100, reg. and menthol

Galaxy king size

Kent king size

Kool Super Longs 100 menthol

L & M king size

Lark king size

Lark Lights king size

Lucky Strike 100

Newport Red king size (hard pack)

Players 100 (hard pack), reg. and menthol

Raleigh king size

Saratoga 120, reg. and menthol

Satin 100

Silva Thins 100, reg. and menthol

Tareyton king size

Tareyton 100

Viceroy king size

Viceroy Super long 100

Virginia Slims 100, reg. and menthol

12–16 mg. tar, 1.0 mg. nicotine, 11–17 mg. CO

Benson & Hedges 100, reg. and menthol

Kent 100

Kool king size menthol

L & M 100

Lark 100

Lark Lights

Marboro king size, reg. and menthol

Marboro 100, reg. and menthol

Montclair king size menthol

Newport Red, king size

Pall Mall Light 100 menthol

Picayune reg. size

Raleigh 100

Richland king size menthol

14–19 mg. tar, 1.1 mg. nicotine, 11–18 mg. CO

Benson & Hedges king size (hard pack)

Camel king size

Eve Lights 120, reg. and menthol

Kool king size menthol (hard pack)

Newport king size menthol (hard pack)

Pall Mall king size

Pall Mall 100

Philip Morris International 100

Philip Morris International 100 menthol

Richland king size

Salem king size menthol

Spring 100 menthol

St. Moritz 100, reg. and menthol

Winston king size

Winston International 100

15–25 mg. tar, 1.2–1.4 mg. nicotine, 11–20 mg. CO	
Chesterfield reg. size	Old Gold Filter king size
Kent 100 menthol	Pall Mall king size
Kool reg. size menthol	Tall 120 menthol
More 120, reg. and menthol	Camel reg. size, non-filter
Newport king size menthol	Lucky Strike reg. size, non-filter
Philip Morris reg. size	Max 120, reg. and menthol
Salem 100 menthol	Raleigh king size, non-filter
Half & Half king size, filter	

19–26 mg. tar, 1.5–1.6 mg. nicotine, 13–20 mg. CO	
Chesterfield king size, non-filter	Tall 120
Herbert Tareyton king size, non-filter	Old Gold Straight king size, non-filter
Newport 100 menthol	Philip Morris Commander king size, non-filter
Old Gold Filter 100	

23–28 mg. tar, 1.8–2.1 mg. nicotine, 11–23 mg. CO	
"Bull" Durham king size	Players reg. size, non-filter
English Ovals reg. size, non-filter	English Ovals king size, non-filter

Source: American Lung Association. Data summarized from a Federal Trade Commission report, 1985: "Tar," Nicotine and Carbon Monoxide of the Smoke of 207 Varieties of Domestic Cigarettes.

Obviously, things may have changed since this study was done, but it does serve to place a brand or variety about where it belongs. It's also interesting to note that "light" cigarettes may contain anywhere from 0.4 mg. nicotine, 4 mg. tar, and 4 mg. carbon monoxide to 1.1 mg. nicotine, 14 mg. tar, and 11 mg. carbon monoxide.

CIRCLES

It's not a bad idea to know your way around the circle, since so many essentials in life are based on it—clocks, angles, maps, compasses, to name just a few. Fortunately, the circle is elegantly simple—a gentle line without beginning or end that is an equal distance from the center at all points. The fact that the circle is divided into 360 degrees is common knowledge, but who did the dividing? And why 360?

The same people who invented the wheel five or six thousand years ago (the Mesopotamians) also fancied the number sixty, basing their entire number system on it. No one seems to know exactly why—it might even have been related to trade: for commercial purposes, it's helpful to use a number that has many factors (see **Commercial Items**, page 40), and sixty has twelve of them, more than any other manageable number—it can be evenly divided by 1, 2, 3, 4, 5, 6, 10, 12, 15, 20, 30, and 60, very useful for marketing. Whatever the reason, the Babylonians picked the system up and passed it on to the Egyptians, who used it to divide the circle into 360 degrees (60 × 6), and who also gave us the symbol (°) for degrees. The Egyptians divided spheres into 360° as well, assigning the first latitude and longitude lines to the earth (see **Latitude and Longitude,** page 107).

This ancient system for dividing the circle has persisted to the present. Each degree (°) of the circle is divided into 60 minutes (60′) and each minute into 60 seconds (60″), divisions which also apply to coordinates for places, compass directions, angles, and, of course, the minutes and seconds of time.

CIRCLE BY DEGREES

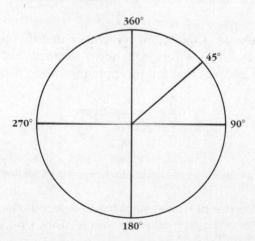

EASY as PI At first glance, measuring the circumference (distance around the edge) and the area inside the circle appears to defy familiar, unbending measuring tools, like rulers. And it would, actually, except for a magical number, discovered about 4,000 years ago, that makes it a piece of cake—or, more to the point, pi.

Pi, symbolized by the Greek letter π, is simply the number of times the diameter of a circle will go around the edge—always 3.1416 times (pi has been calculated to something like a billion places, but most people stop at four). This means that you *can* use your ruler to measure the circle—all you need to measure is the diameter or the radius (half the diameter).

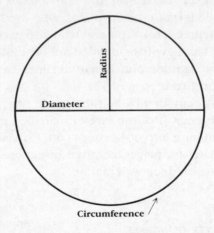

If the diameter is 6", the radius is 3" and the circumference 18.85".

The formulas are simple, but remember that if you measure in feet or inches, the answers will be in decimal feet or decimal inches. The examples that follow apply to a circle with a diameter of 6 inches.

$$d\pi = \text{circumference}$$
(6 inches times 3.1416 = 18.85 inches)
OR
$$2\pi r = C$$
(2 times 3.1416 times 2 inches = 18.85 inches)

You can also use pi to find the area in square inches (or square centimeters or whatever measure you're using) by squaring the radius (multiplying it by itself) and multiplying that times pi (the example applies to the above circle, which has a radius of 3 inches):

$$\pi r^2 = \text{Area}$$
$$(3.1416 \text{ times } 9 \ [3 \text{ inches squared}] = 28.27 \text{ square inches})$$

So how do you remember a number like pi? A close approximation is 22/7. Or look at your calculator—most calculators have the symbol for pi on one of the buttons. Push it and pi will appear.

CLOTHING SIZES

The sizes for most American clothing—men's, women's, and children's—are based on *anthropometric studies,* which measured the size of the varying human, in this case American, form. The first anthropometric studies used by the clothing industry for women's as well as men's apparel were conducted during World War II by the U.S. Armed Forces, which had to measure their burgeoning ranks for uniforms. Another study, measuring 150,000 children in sixteen states, was done by the Department of Agriculture at about the same time. American clothing sizes, even women's sizes, are still frequently based on these studies.

The clothing industry has also introduced the Small-Medium-Large sizes (S-M-L), sometimes adding an Extra-Small (XS) and an Extra-Large (XL), which can apply to almost any article of clothing on the market. To figure out how the system works, you can safely assume that each letter represents two, sometimes three, numbered sizes, spreading over the size range. There is much license taken here, however; one company's Large might be another's Medium, especially if the first was made in the Far East, where the anthropometric measurements tend to be smaller.

CLOTHING, BABIES' What could be more frustrating than trying to buy clothing for somebody else's baby? If you've never purchased baby clothes before, you could easily blunder, for the sizes are misleading. The confusion begins with three sizing systems:

The age system may include the following sizes: newborn, 3 months, 6 months, 12 months, 18 months, 24 months, and 36 months. Motherly wisdom has it that you multiply the child's age by two—assuming the infant is not a preemie or especially large for its age—in order to get the right fit (for example, a 6-month-old baby would wear a 12-months size). When the child is around 12 months, however, you buy the *Toddler* sizes, which begin with the year-old baby and end with the 4-year-old: 1T to 4T. Sizes followed by a T are supposed to have extra room for diapers (although T may sometimes mean Tall).

The weight system is supposed to apply to the following ages: up to 14 pounds (newborn to 3 months), 15 to 20 pounds (up to 6 months), 21 to 26 pounds (up to 18 months), 27 to 32 pounds (up to 24 months), etc., but brands vary.

The newborn to extra-large system is anybody's guess, for there are no government rules—only suggested guidelines—in the industry to encourage consistency. There is usually a weight or age on the label for guidance.

Choosing a size wouldn't be half so difficult if these systems were accurate—if a size 18-month sleeper actually fit an 18-month-old baby, or overalls for a 15 to 20 pounder fit your 19-pound nephew. Often, however, they turn out to be too small. Perhaps babies are bigger than they used to be, or the clothes shrink with all the washings. Whatever the reason, what are you to do? Buy gifts one or two sizes bigger than what seems perfectly sensible. Babies grow fast—if the cuties don't fit into them tomorrow, they probably will next week.

CLOTHING, CHILDREN'S A baby is considered by the clothing industry to have become a child when he or she can tear around independently and diaperless. Children's clothing sizes apply to both girls and boys, and, unfortunately, are no more dependable than infant and toddler sizes. Children's sizes continue to estimate age—2 through 6X (X meaning extra large). After size 6X, children experience one of the American rites of passage: girls and boys split up to find their finery in different departments.

Girls' sizes continue the children's sizes, using only even numbers after 7—8, 10, 12, 14, and 16. This is confusing because the most popular women's size numbers are exactly the same! But although a girl's 16 fits one girl weighing between 99 and 112 pounds, it might take two girls of that weight to fill out a woman's 16. Unlike women's sizes, girls' clothing may also be labeled for width—slim, regular, and chubby. A teenage girl graduates to the *Junior Department* (see **Clothing, Women's,** page 36).

Boys' sizes, like girls' sizes, are based on age, even though it rarely applies. The size range is from 6 to 24, using only the even numbers (6, 8, 10, etc.). Boys' clothes are usually labeled for width as well—slim, regular, and husky. The larger boys' clothes are often in a separate department for "young men," but the sizes sensibly continue the same system (unlike the odd-numbered Junior clothes for teenage girls). A teenage boy who can wear the larger boys' sizes can probably fit into men's sizes as well—the overlap is considerable. Young men's clothes are often cut slimmer, however, and the styles are more likely to reflect the current fads.

CLOTHING, MEN'S No size system is so sensible as that applied to men's clothing. It is an island of reason in a sea of size insanity (see other **Clothing** entries). For one thing, the numbers that represent men's clothing sizes usually mean inches. For another, because sizes are actual measurements, they are consistent—a size 15 shirt's collar band measures 15 inches for any brand. Men are lucky for a third reason—better shirts and pants are often measured in two directions, so alterations are frequently unnecessary.

Men's shirts and sweaters are often sized to fit a man's neck and arms, the measurements applying to the collar band and sleeve length (measuring from the base of the collar, over the shoulder to the edge of the cuff). Most collar band sizes increase every half-inch from 14 to 16½. The sleeve length usually measures from 32 to 36 inches.

Men's pants have two measurements: waist and inseam (crotch to hem measurement), so a short, thin man will not have to have a tall, thin man's pants shortened. (Women are not so lucky.) Waist sizes run from 29 to 39, inseams from about 29 to 35 inches.

Men's coats and jackets (and sometimes sweaters) are

sized by chest measurements in inches, going from 32 to 44 and higher.

The only monkey wrench in this system is the use of Extra Small (XS), Small (S), Medium (M), Large (L), and Extra Large (XL) (see page 33). These sizes most often apply to men's sportswear and underwear, but sometimes to shirts and sweaters as well, and they do vary some from brand to brand. The fit may not be as precise as measured sizes—in a Small, all measurements will be small, while a Large will have all large measurements, with Medium somewhere in between.

CLOTHING, WOMEN'S One of the more outrageous systems foisted upon American women is the sizing of their clothes. Women's clothing is not sized to the last half-inch as many men's clothes are, not only because the huge variety of designs sometimes makes exact measurements difficult, but because the women's clothing industry seems reluctant to confront a woman with her own measurements. But if measurements can't be used, then how to label clothes for the vast variety of heights, girths, and shapes represented by American women?

The clothing industry has dealt with this problem with this solution: Instead of dividing women's clothing into two groups—teenagers' clothing and adult clothing—it has divided women's apparel into no less than five main groups. Many women who flounder between one or another department don't realize that the real dividing factor is *height and figure type.* Here are some general, if unavoidably unreliable, guidelines:

Junior:	Sizes: 3-5-7-9-11-13-15
	Height: approximately 5'2" to 5'5"
	Figure type: slender
Misses:	Sizes: 4-6-8-10-12-14-16-18
	Height: approximately 5'5" to 5'7"
	Figure type: developed and well proportioned
Misses Petite:	Sizes: 2P-4P-6P-8P-10P-12P-14P-16P
	Height: approximately 4'8" to 5'4"
	Figure type: developed and well proportioned
Half:	Sizes: even numbers from 10½ to 26½
	Height: approximately 5'2" to 5'4"
	Figure type: fuller and rounder than Misses

Women's: Sizes: even numbers 34 to 52
             Height: approximately 5'5" to 5'8"
             Figure type: fuller and rounder than Misses

Based on data from Figure Types and Size Ranges, *by Debbie Ann Gioello, Courtesy of Fairchild Books, a division of Fairchild Publications, New York.*

Not one of the size types deals with a woman tall enough to be a model, whose lithe length every American woman is supposed to be striving to resemble. And it does sound as if the sizes assigned to women (34 to 52) were bust measurements, but do not be taken in by this: a size 34 fits a woman with a 38-inch bust and hips and a 29-inch waist.

Size-foolery is also used by the makers of "better" clothes—although there is general agreement in the clothing industry about size ranges, a woman who wears a size 14 in clothes she can afford frequently will fit nicely into a more expensive 12 or 10, a practice that many fashion experts seem to feel persuades the woman to buy pricier clothes.

Styles vary greatly between size types as well. Usually the bigger a woman is, the less trendy and exciting have been her style choices. This dreadful practice has incensed full-bodied women, so today special stores and departments are offering a more tempting variety.

COMFORT INDEX (WEATHER)

Weather reports often include figures that are not actually the weather, but the weather's effects on the human body. In winter the wind can make lower temperatures feel colder. In summer the humidity can make high temperatures feel hotter. In either case the effects on the human body can be acutely, even dangerously, uncomfortable.

Most people are familiar with the *wind chill factor*—a day with a temperature of 20 degrees Fahrenheit and a wind speed of 30 mph will be reported as having a wind chill factor of −2 degrees. This does not mean that the wind makes the air colder. It does mean

WIND CHILL CHART

To determine wind chill, find the outside air temperature on the top line, then read down the column to the measured wind speed. When the outside air temperature is 0° and the wind speed is 20 miles per hour, for example, the rate of heat loss is equivalent to minus 39° under calm wind conditions.

Equivalent Temperature (°F)

Wind Speed (miles per hour)	35	30	25	20	15	10	5	0	−5	−10	−15	−20	−25	−30	−35	−40	−45
Calm	35	30	25	20	15	10	5	0	−5	−10	−15	−20	−25	−30	−35	−40	−45
5	32	27	22	16	11	6	0	−5	−10	−15	−21	−26	−31	−36	−42	−47	−52
10	22	16	10	3	−3	−9	−15	−22	−27	−34	−40	−46	−52	−58	−64	−71	−77
15	16	9	2	−5	−11	−18	−25	−31	−38	−45	−51	−58	−65	−72	−78	−85	−92
20	12	4	−3	−10	−17	−24	−31	−39	−46	−53	−60	−67	−74	−81	−88	−95	−103
25	8	1	−7	−15	−22	−29	−36	−44	−51	−59	−66	−74	−81	−88	−96	−103	−110
30	6	−2	−10	−18	−25	−33	−41	−49	−56	−64	−71	−79	−86	−93	−101	−109	−116
35	4	−4	−12	−20	−27	−35	−43	−52	−58	−67	−74	−82	−89	−97	−105	−113	−120
40	3	−5	−13	−21	−29	−37	−45	−53	−60	−69	−76	−84	−92	−100	−107	−115	−123
45	2	−6	−14	−22	−30	−38	−46	−54	−62	−70	−78	−85	−93	−102	−109	−117	−125

COLD
VERY COLD
BITTER COLD
EXTREME COLD

"Calm-air" as used in wind-chill determinations actually refers to the conditions created by a person walking briskly (at 4 miles per hour) under calm wind conditions.

Source: National Weather Service, U.S. Department of Commerce.

HEAT INDEX CHART

Air Temperature and Relative Humidity versus Apparent Temperature

Heat Index (or Apparent Temperature)

AIR TEMPERATURE (°F)	RELATIVE HUMIDITY (%)																				
	0	5	10	15	20	25	30	35	40	45	50	55	60	65	70	75	80	85	90	95	100
140	125																				
135	120	128																			
130	117	122	131																		
125	111	116	123	131																	
120	107	111	116	123	130	139	148														
115	103	107	111	115	120	127	135	143	151												
110	99	102	105	108	112	117	123	130	137	143	150										
105	95	97	100	102	105	109	113	118	123	129	135	142	149								
100	91	93	95	97	99	101	104	107	110	115	120	126	132	138	144						
95	87	88	90	91	93	94	96	98	101	104	107	110	114	119	124	130	136				
90	83	84	85	86	87	88	90	91	93	95	96	98	100	102	106	109	113	117	122		
85	78	79	80	81	82	83	84	85	86	87	88	89	90	91	93	95	97	99	102	105	108
80	73	74	75	76	77	78	79	79	79	80	81	81	82	83	85	86	86	87	88	89	91
75	69	69	70	71	72	72	73	73	74	74	75	75	76	76	77	77	78	78	79	79	80
70	64	64	65	65	66	66	67	67	68	68	69	69	70	70	70	70	71	71	71	71	72

Source: National Oceanic and Atmospheric Administration, U.S. Department of Commerce.

that it makes *people* colder—that an appropriately bundled-up person's body will react to that wind and temperature *as if it were* −2 degrees. Although you lose one-fifth of your body heat by breathing, you can control some of the heat loss with extra layers of well-insulated clothing. The effects also vary some with the size of the person—small persons lose heat faster than large ones.

It's important to know the wind chill factor on particularly cold and windy days, to avoid *hypothermia,* or dangerously low body temperatures. On a dry winter day the cold can sneak up on you. Use the wind chill chart on page 38 for guidance.

Perspiring is the human body's cooling system, and it works quite well in summer as long as the air is dry. Damp air doesn't allow good evaporation, however, so the humidity can really bother you—pleasant temperatures in the low 80s can feel oppressive if the relative humidity is over 70 percent. In fact, the average person can be outside in a 100-degree temperature with 90 percent relative humidity for only ten minutes. The effects of humidity and temperature on the human body is reported as the *humiture,* or sometimes as the *heat stress index.* When the heat stress index is high, outdoor activities can be dangerous to your health. See the heat index chart on page 39.

COMMERCIAL ITEMS

Why is it that our dollar is based on the number ten (10 × 10 = 100), but the things we buy with it are sold by the dozen? When the contemporary trend is toward the ten-based metric system, why do manufacturers and commercial establishments still frequently base their trade on the number twelve?

Actually, when it comes to selling and pricing things, twelve is a much more useful number than ten, because it has more factors (divisors). The number ten only has four divisors—1, 2, 5, and 10—while twelve can be divided by 1, 2, 3, 4, 6, and 12, making it much more flexible. This convenience may partially explain why the number twelve has been popular for hundreds of years, with twelve inches making up a foot, and twelve ounces the troy pound (see **Weight,** page 203).

Science writer Isaac Asimov, noting these advantages in his book *Realm of Numbers,* suggests that if humans had been born with six fingers on each hand instead of five, our number system would likely be based on twelve instead of ten!

Commercial items are counted in multiples of twelve, as follows:

12 items = 1 dozen

12 dozen = 1 gross

12 gross = 1 great gross (1,728 items)

COMPASS

On the face of it, the compass seems unbelievably simple and handy—it's affordable, portable, and its little magnetic needle always points to magnetic north. It's even easier to read than it used to be. Before 1920 there were 32 bewildering direction points—north, north by east, north-northeast, northeast, northeast by east, east-northeast, east by north, east, to name only eight.

Today you simply read clockwise from zero to 360—north is 0 degrees, east is 90 degrees, south is 180 degrees, and west is 270 degrees. (Like the circle, the compass is divided into 360 degrees.)

COMPASS

Source: Thermometer Corporation of America.

If you're lost in the woods and you remembered your compass, you probably assume that all you have to do is line up north on the compass with the needle, and you'll get an accurate read of all the other directions. Unfortunately, you also need to consider the whereabouts of magnetic north.

Magnetic north is not the same as true north, but you probably have assumed that it's pretty close. Even if you know that magnetic north moves around a bit, you probably envision it moving around the North Pole. But magnetic north is located about *1,200 miles* from the North Pole, in the vicinity of Prince of Wales Island, which is off the Canadian coast about halfway between the U.S. border and Alaska! You can see how this eccentric location can seriously throw off a compass reading—if you're lost in Alaska or the Yukon, for example, your compass won't point anywhere near north; it might, in fact, point south. Even in the "Lower Forty-nine," the only way to get an accurate reading from your compass is to know the number of degrees you need to correct for—a correction known as the *declination.*

You'd think that this sort of essential information would be easy to find, but it isn't. For one thing, the declination—how many

MAGNETIC DECLINATION IN THE U.S., 1985

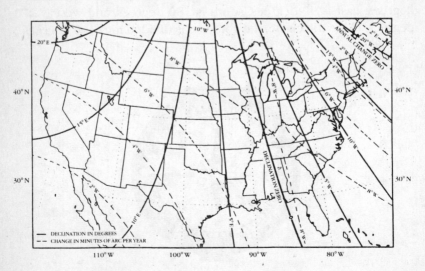

Source: U.S. Geological Survey, Department of the Interior.

degrees to add or subtract to find true north—changes about ¹/₁₅ of a degree every year—not serious, but old maps can be off a degree or more. For another, although declination information is always included on topographical maps and boating charts, it's hard to find anywhere else. The public library might be able to tell you the declination for your area, but unless they have a topographical map collection, they'd probably end up calling one of the National Cartographic Information Centers or the U.S. Geological Survey office, often a long-distance call. Sometimes you can get the information from mountaineering, boating, or sporting goods stores that sell "topo" maps and charts.

How essential this information is to you depends on where you plan to do your adventuring—as you can see on this rough estimate of declinations, it's those of you in western and northeastern states whose compasses will be farthest off. If you're lucky, you'll be outdooring in more centrally located states where there's only a 5 degree or less correction, probably okay in a pinch: Wisconsin, Illinois, Michigan, Ohio, West Virginia, Indiana, Kentucky, Tennessee, western North Carolina, South Carolina, Georgia, Alabama, most of Mississippi, and Florida.

Authorities at the National Cartographic Information Center recommend finding true north the old-fashioned way—by locating the North Star (the very bright star in a direct line with the top star on the front edge of the Big Dipper) because it never deviates from true north more than one degree. If you line up "north" on your compass with the North Star, you will see how many degrees' difference there is between true north and magnetic north, to which the needle is pointing.

The intent of all this information is not to instruct you in compass finesse, but simply to warn you about making dangerous assumptions about its accuracy. Here's an interesting footnote: the needles on the earliest compasses, used by the Chinese about 1000 A.D., were marked to point *south*!

COMPUTERS

What's the difference between a 128 Macintosh and a 512? It's K— or kilobytes—meaning units of memory. This may have no meaning to you at all; a more descriptive term would have been helpful. The computer people have redefined so many familiar words, even metric prefixes that are practically sacred, that you hardly dare trust yourself when you're around them. What follows is a brief introduction.

Generally, as personal computers (known as PCs) have grown, so have their memory storage capacities. A unit of memory in a computer is equal to a *byte,* which can store one character of text, such as a "b," a comma, or a space. Each byte is made up of an 8-digit yes-no (1-0) code, and each of those digits is called, just to confuse you, a *bit,* short for *binary digit.* A kilobyte, or 1K, consists of 1024 bytes; 1,048,576 bytes (or 1,000K) is a *megabyte. Kilo-* usually means thousand, and *mega-* usually means million, except in computers, where they have been redefined to deal with a different number system: Computers work naturally in powers of 2 rather than in powers of 10. When you push each system—the 2-based and the 10-based—to a million, this is what you get:

Powers of 2		Powers of 10	
2		10	
4		100	
8		1000	= 1 K
16		10,000	
32		100,000	
64		1,000,000	= 1 M
128	= 1 K		
256			
512			
1,024			
2,048	= 1 M		
4,096			
8,192			
16,384			
32,768			
65,536			
131,072			
262,144			
524,288			
1,048,576			

To summarize:

COMPUTER MEMORY UNITS
8 bits = 1 character = 1 byte
1000 bytes = 1 kilobyte (K)
1,000,000 bytes = 1,000 kilobytes = 1 megabyte

How do these figures translate for general use? Computers store text using 1 byte per character, including spaces, which the computer reads as character. A typical typed, double-spaced page of about 350 words is about 2,000 characters, requiring 2K of computer memory or 2K disk space per page. Drawings, charts, and other complexities manageable by most computers add more Ks, however—how many can be difficult to judge. How much material you can store on your disks is measured in Ks—400K is common for a single-sided disk and 800K for a double-sided disk. Finally, a computer has two kinds of memory, and the capacity of each is measured in Ks.

COMPUTER MEMORY The subject of computer memory is so complicated that for every simplified explanation there are impressive exceptions. A beginner needs a few generalizations, however, to gain a basic understanding of this material. To start, here are a few important points:

The two kinds of computer memory have such similar names that the uninitiated have a hard time keeping them apart: ROM and RAM. ROM (Read Only Memory) is something like the computer's brain, which directs the computer's functions. You normally do not alter the contents of ROM, and the contents remain the same whether the power is on or off.

The contents of RAM are lost when the power goes off. The RAM (Random Access Memory) is also called auxiliary storage—when a computer is said to have 128K of memory, unless otherwise noted, this almost always means RAM. RAM lets you use programs like a word processing program, games, data base programs, etc. (depending on which disk you have inserted), stores your material, accommodates your changes and, at your command, will save your material. You need enough memory space (K) in RAM to read the programs most useful to you, with room left for other functions. Some computers, for example, store the system (basic computer function instructions) in ROM, while others require RAM space for this.

There was a time when personal computers had such tiny RAMs—64K and even 16K for some game models—that calculating your RAM needs accurately was extremely important. Today personal computers are becoming so powerful that a megabyte of RAM is not uncommon. This much RAM space is probably adequate for most ordinary needs. The author wrote this book on a computer with 512K of RAM, storing the manuscript plus notes, bibliography, and a small data base program on one 3½-inch 800K disk, and the system (basic computer function instructions) plus a word processing program on another. She worked with the two disks simultaneously by using two disk drives—or disk "readers." (The RAM doesn't need 1600K to deal with two 800K disks—it simply goes to the disk and reads what it needs when it requires it, much as you find a definition without reading the entire dictionary.)

CONSUMER PRICE INDEX (CPI)

Many people are looking for the actual cost of living when they consult the Consumer Price Index, not an unreasonable mistake, since the CPI used to be called the "Cost of Living Index." But if you're looking for cost figures in dollars, the figures make no sense. Still, the Consumer Price Index is a statistical paradise that is relatively easy to understand and use, so don't be put off; instead glance quickly over this typical monthly report—this one is for the California area, one of ten such regions—easily found even in small public libraries or reported monthly on newspaper business pages. Then read on.

The first confusing item is the notation under the heading *All Items Indexes,* which reads "1982–84 = 100 unless otherwise noted." This is the key to the Consumer Price Index, for in reading the CPI, it is essential to remember that *the CPI reports are not the*

Beginning with the release of this data for January 1988, the standard reference base for the Consumer Price Indexes is 1982–84 = 100. *As a convenience to users, we will continue to publish all items indexes on their former official reference base (1967 = 100 in most cases).*

CONSUMER PRICE INDEXES
Pacific Cities and U.S. City Average

ALL ITEMS INDEXES *(1982–84 = 100 unless otherwise noted)*
JANUARY 1988 AND 2ND HALF 1987

MONTHLY DATA

ALL URBAN CONSUMERS

	INDEXES			PERCENT CHANGE		
				Year ending		1 Month ending
	JAN. 1987	DEC. 1987	JAN. 1988	DEC. 1987	JAN. 1988	JAN. 1988
U.S. City Average	111.2	115.4	115.7	4.4	4.0	0.3
(1967=100)	333.1	345.7	346.7	—	—	—
Los Angeles-Anaheim-Riverside	113.4	118.5	118.9	5.1	4.9	0.3
(1967=100)	335.1	350.2	351.2	—	—	—
San Francisco-Oakland-San Jose	112.5	117.4	118.4	5.0	5.2	0.9
(1967=100)	345.8	360.9	363.9	—	—	—

URBAN WAGE EARNERS AND CLERICAL WORKERS

	INDEXES			PERCENT CHANGE		
				Year ending		1 Month ending
	JAN. 1987	DEC. 1987	JAN. 1988	DEC. 1987	JAN. 1988	JAN. 1988
U.S. City Average	110.0	114.2	114.5	4.5	4.1	0.3
(1967=100)	327.7	340.2	341.0	—	—	—
Los Angeles-Anaheim-Riverside	110.8	115.7	115.9	5.1	4.6	0.2
(1967=100)	327.4	342.0	342.7	—	—	—
San Francisco-Oakland-San Jose	111.3	116.4	117.5	5.1	5.6	0.9
(1967=100)	335.0	354.4	357.7	—	—	—

Source: U.S. Department of Labor, Bureau of Labor Statistics.

prices themselves, but the percent that prices have changed com-pared to an earlier time. Before 1988 this earlier time, called a base period, was 1967. The base period always equals 100 (not zero, as one might think). Today, 1982–84 is the base period. An increase of 203 percent since 1982–84 would be shown in the index as 303.00, or 203 more than 100. This looks odd, since most people don't think of percentages over 100 percent, but it makes it easy to translate into dollars. If a "market basket" full of goods and services cost $10.00 in 1982–84, the same items would cost $30.30 when the index reads 303.00 points. (If the base has changed again, not to worry; it is always specified.)

The second confusing item is that there are *two* CPIs, one for *All Urban Consumers* and one for *Urban Wage Earners and Clerical Workers.* This results in two reports—one on the left and one on the right—that look the same but contain different figures. The CPI is used to measure inflation generally, but also to measure the cost of living for wage earners negotiating new contracts. The average "market basket" of wage earners and clerical workers is a little different from the general average—one group may not spend as much on an item as another. It's just a matter of "weighting"—the Bureau of Labor Statistics (U.S. Department of Labor) has to know what percent of the average budget is spent on an item before it can determine the cost of living for each group. And the numbers are really interesting! Do you want to know how much of your budget you spend on bananas? The Bureau of Labor Statistics thinks the average for All Urban Consumers is .088 percent. You also average .251 percent on coffee, .260 percent on sofas, .918 percent on footwear, .323 percent on newspapers, 4.8 percent on gasoline, 2.181 percent on the telephone, and .767 percent on the dentist. These are just a few of the 364 items that the Bureau of Labor Statistics (BLS) gathers data on from about 21,000 retail establish-ments and 60,000 housing units in 91 urban areas. (It doesn't cover rural areas.)

So now you can read that January 1988 CPI:

1. The first item is the U.S. City Average for *All Urban Consum-ers.* It compares the January 1988 figures (3rd column) to the month before (column 2) and a year before (column 1). The next three columns calculate the percent change for the same periods.

2. The next two sets of similar figures are for the large cities in the region, in this case, the L.A. area and the San Francisco area.

3. Read the lower portion of the index—*Urban Wage Earners and Clerical Workers*—the same way, but apply the information to a more select group.

For some really fascinating reading, next time you're in the library ask for *Consumer Price Index Detailed Report,* also a monthly publication, which gives the percent of increase (or decrease) in prices for each specific category—like those banana, footwear, and other items mentioned earlier. The figures read the same way.

There's one other confusion: CPI data are often noted as being "seasonally unadjusted" or "seasonally adjusted." Because price data are used for different purposes, the CPI publishes both sets of data in their *Detailed Report.* The "seasonally unadjusted" figures are the exact figures for that period. The "seasonally adjusted" figures make allowances for price changes affected by climate, holidays, etc., so they reflect the true overall price; e.g., prices that go up at Christmas often aren't pushed up by inflation but by the holiday demands, and will go down again after the holiday. If you're using the index to show the rate of inflation, predictable price inflators like holidays have to be considered.

Do you want to know what percent more you're paying for fresh whole chickens now than you did in 1982–84? Or what percentage gasoline has increased in the last year? Look it up on the *CPI Detailed Report.* The *CPI Detailed Report* also gives figures for many more cities than found on the card, although not all are reported on monthly. The next time you see CPI figures in the paper, you'll be able to figure the rate of inflation for yourself.

COPYRIGHT PAGE

If you need to know when a book was published, you look on the flipside of the page announcing the title, author, and publisher.

There you will find the copyright date, usually preceded by a cir-
cled "c." That's clear enough. But what are all those other numbers
for?

LIBRARY OF CONGRESS CATALOGING IN PUBLICATION DATA
Blocksma, Mary.
Reading the numbers : a survival guide to the measurements,
numbers, and sizes encountered in everyday life / Mary
Blocksma.
p. cm.
ISBN 0 14 01.0654 5
1. Weights and measures.   2. Weights and measures—United States.
I. Title.
QC88.B54   1989
530.8—dc19   88–23219

Most of the remaining numbers are included under the Library of
Congress heading for librarians ordering and cataloging books.

QC88.B54   is the Library of Congress classification number.
           Every book published by a United States publisher is
           classified by the Library of Congress.
   1989   is the year the book was published.
  530.8   is the suggested Dewey Decimal classification for
           smaller libraries.
88-23219   is the number used by librarians to order catalog cards.
    ISBN   0 14 01.0654 5 (See **ISBN Numbers,** page 105.)

A short publishing history is often found at the bottom of the
page—a strange string of numbers that looks something like this:

1   3   5   7   9   10   8   6   4   2

Publishers use these numbers to keep track of the book's printing
history without having to reset the page with each new printing—
they simply remove the numbers that no longer apply. Read the

lowest number for the number of the printing. This book is in its first printing.

CRASH TEST RATING INDEX (CTRI)

Range: About 1,200 (safest car) to 4,500 (least safe car)

How do you know if one new car is safer than another? At last there is a number called the CTRI, or the Crash Test Rating Index. The Crash Test Rating Index score is assigned by the U.S. Department of Transportation to all new car models, which are crashed into a concrete barrier at 35 miles per hour (similar to two cars crashing head-on at 35 miles per hour). Models with the least damage to driver and car are assigned lower numbers than those suffering more damage. The numbers range from about 1,200 for the safest car to 4,500 for the least safe car.

Any car dealer should have these test numbers, which are also available in *The Car Book* by Jack Gillis (Harper & Row, updated annually), found in most public libraries.

CURRENCY (NOTES)

Nearly all the U.S. "greenbacks" used today are Federal Reserve Notes, produced by the Bureau of Engraving and Printing in Washington, D.C., and now available only in denominations of $1, $2, $5, $10, $20, $50, and $100. There was a time, however, when you could get $500, $1,000, $5,000, and $10,000 notes. Although the Federal Reserve stopped printing those huge denominations in 1945, they weren't withdrawn from circulation until 1969. The actual size of currency today is also smaller—until 1929, bills measured 7.42 × 3.13 inches, which was fairly large compared to today's 6.14 × 2.61 inches.

What do the numbers mean on the face of today's currency?

NUMBERS ON A TEN-DOLLAR BILL

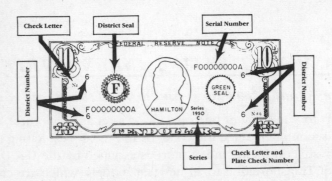

Source: U.S. Federal Reserve Bank of Atlanta.

The serial number, which appears in both the upper right and lower left portion of the bill, has eight digits, with a letter preceding and following it, each considered part of the serial number. The prefix letter is the same as the one in the bank seal, telling you which of the twelve Federal Reserve Banks (the United States is divided into twelve Federal Reserve Districts) issued the note. Each Federal Reserve District issues its own Federal Reserve Notes according to its region's needs. The Federal Reserve District Number, printed in four places on the bill, is simply a number translation of the letter:

FEDERAL RESERVE DISTRICTS

1	Boston	A	7	Chicago	G
2	New York	B	8	St. Louis	H
3	Philadelphia	C	9	Minneapolis	I
4	Cleveland	D	10	Kansas City	J
5	Richmond	E	11	Dallas	K
6	Atlanta	F	12	San Francisco	L

Locate your Federal Reserve District on the map; then check your currency. Most of the bills in your pocket probably carry the same bank seal letter—they were issued by your Federal Reserve District.

MAP OF FEDERAL RESERVE DISTRICTS

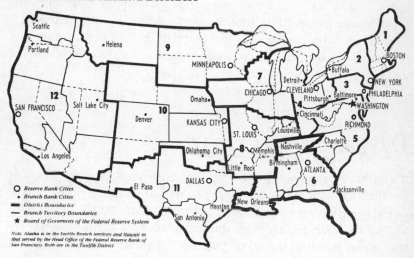

Source: U.S. Federal Reserve Bank of Atlanta.

The notes are numbered in lots of one hundred million, with a "star" note substituted for the one-hundred-millionth note. The first run will carry the suffix letter A, running through Z for succeeding runs (but omitting O so it won't get mixed up with zero). When you consider that there are several billion $1 notes alone in circulation, it is amazing to realize that no two notes of the same kind, denomination, and series have the same serial number!

Star Notes. If a note is damaged during printing, interrupting the number series, it is replaced with a "star note," which is exactly like it except that a star takes the place of the suffix letter; e.g., F00000004*.

Series Number. The year the design of the note was first used is shown as the series number, next to the signature of the Secretary of the Treasury in the lower right-hand corner. If only a minor change is made, one not requiring a new engraving plate, a letter will appear after the year, starting with A for the first change in that series; e.g., 1985B, indicating that two minor changes were made.

Note Check Letter and Plate Serial Number. In the upper left-hand corner is a small capital letter followed by a number. This is the Note Position, or Check Letter, indicating the position of the note on the printing plate. The newer presses print thirty-two notes to a sheet. The same letter appears near the bottom right-hand corner, followed by another number, which is the Plate Serial Number, or the number of the plate from which the note was printed. A 46, for example, would be the forty-sixth plate made for that type, denomination, and series of note.

DISTANCE, NAUTICAL

If you have ever been put off by that outlandish language used at sea then nautical measures are probably not your cup of tea. There doesn't seem to be any good reason for sailors to talk differently and measure differently from the rest of us. Still, while that may be true about much of the sailor talk, the measures—those knots and nautical miles and such—actually do make sense.

A look at the origins of the mile we use on land illustrates this. The Romans measured the first land mile as 1,000 paces (*milia passuum* in Latin, hence our word "mile"), with each pace equaling two marching steps. This mile, at 1,618 of our yards, was a little short of today's mile. After centuries of variation, Elizabeth I assigned the mile today's familiar 1,760 yards, or 5,280 feet, calling it a *statute mile,* a term that has endured with the length. Walking to establish distance, however, just doesn't work on water, and the nautical mile had been around long before the Romans came along, anyway.

The *nautical mile* is not based on human measures but on the circumference of the earth, which has for centuries been divided into 360 degrees, with each degree divided into 60 minutes. Ancient Chaldean astronomers are believed to have established the first nautical mile based on degrees before 600 B.C., and for most of history sailors have charted their course, progress, and whereabouts by degrees of latitude and (later) longitude (see **Latitude and Longitude,** page 107). To measure distance by degree made sense. Like other measures, however, the actual length of the nautical mile varied a great deal through the centuries, with different countries dividing each degree into 10, 12, 15, or 60 parts.

Even in the twentieth century there has been wide disagreement on the precise measure of the nautical mile. The British and the United States, for example, made the nautical mile equal to a minute ($\frac{1}{60}$) of a degree, but the British nautical mile measured 6,080 British feet, while the U.S. nautical mile measured 6,080.20 U.S. feet (1853.248 meters). The British foot is $\frac{1}{400,000}$ smaller than the U.S. foot, which sounds silly to mention but is actually quite important for precise measurements. In 1929 a large number of countries adopted the International Nautical Mile, defined as 1,852 meters, but the United States, the U.S.S.R., and Great Britain abstained. Only in 1954 did the United States finally adopt the International Nautical Mile of 1,852 meters. Today's nautical measures of distance are as follows:

U.S. NAUTICAL MEASURES OF DISTANCE
6 feet (exactly) = 1 fathom
120 fathoms = 1 cable = 720 feet
8.44 cables = 1 international nautical mile = 1,852 kilometers =
6,076.10333 feet

Knots are simply nautical miles per hour: 1 knot = 1,852 kilometers per hour. Knots and nautical miles are not limited to seafarers—airfarers, including all of today's commercial and private pilots, have been using them for years.

DOW JONES INDUSTRIAL AVERAGES

There must be many millions of people who have wondered for years what that information on the nightly news about how many points up or down the Dow Jones went means. What are points anyway, and what is the Dow Jones the average of?

To further complicate matters, Dow Jones reports three averages: a Transportation Average, a Utilities Average, and an Industrial Average. However, when you hear a "Dow" or "Dow Jones" or "Dow Jones Average" figure, you can safely assume that what is being referred to is the Dow Jones *Industrial* Average—the average cost per share of stock from thirty companies, chosen by the Board of the New York Stock Exchange as the most outstanding companies reflecting the variety of American business, to represent the stock market. And the "points" can be directly translated into dollars and cents.

In January 1987 the Dow Jones Industrial Average went over 2,000 points for the first time in its ninety-one-year history. At the time, its highest-priced stock sold for about $122.85 a share, and its lowest-priced stock was $5.37. In fact, if you added up all thirty stocks, they totaled only $1,780. How could the average have been higher than the sum total? The answer has to do with stock "splits" and "substitutions."

The first Dow Jones Averages report, which appeared in *The Wall Street Journal* in 1896, really was a simple average—Mr. Charles Dow, who devised the system and founded a company with Edward Jones, simply added up the prices of the then twelve stocks and divided by twelve. However, the companies' stocks began rising. To make stock shares affordable to small investors, companies began to split their shares when the prices got too high. (Most stocks are split before they reach $100.)

Finding the average was no longer simple. When a company split its stock, it halved the share size, so it could offer twice the shares at half the price. The market hadn't fallen, but if a simple average was calculated, the Dow Jones would go down. So Dow Jones no longer divided the total stock prices by the number of companies, but by a "divisor," a number that would make the final "average" the same as it would have been before the split.

The first divisor used was 16.67. Over the years, as companies were added to the list and stocks split over and over again, the divisor got lower and lower, until it reached 1. When the average went over 2,000 points, the divisor was .0877, even less than 1, which made it a multiplier. The 2,000 figure represents the price per share of stock *if there had never been any splits or substitutions.*

Despite ups and downs, the Dow Jones Industrial Average has been on the rise. In 1906 the stock market closed at over 100; in November 1972 it first closed over 1,000, and in January 1987, with much hoopla, it passed the 2,000 mark. The Dow Jones shows how the average price per share of stock has risen—shares worth $100 in 1906 might well have sold for $2,000 in 1987.

DWELLINGS (SIZE)

The size of a house is always described in square feet. But how do you know from the real estate ad that caught your eye whether the 1,200-square-foot home with the backyard peach orchard is big enough to warrant a visit? Faced with hundreds of ads, you need some way to sort out the best bets. What does 1,200 square feet mean?

The square-foot description is a measure of the inside finished living space. This includes all the finished floor space in the house, so if there are two floors, the second floor is counted as well. *Not included* are the garage, outside porches, decks, or an unfinished basement.

A house is usually measured by an appraiser hired by a bank or mortgage company to help establish the home's market value. The appraiser works from the *outside,* measuring off a rectangle, then adding the living spaces that lie outside it and subtracting the empty areas inside. The final figure, therefore, includes the space occupied by inside and outside walls. (In some areas your real estate agent should know if this applies to you—it is customary to measure the house from the inside, a tedious procedure that requires measuring every room. This final figure does not include the walls, so it would be somewhat smaller.)

So how big is 1,200 feet? Real estate agents report that you never know for sure until you see the place, but as a rule of thumb, a 1,000-square-foot house is considered small, a 1,500-square-foot house average, while one with more than 2,000 feet is pleasantly large.

Once you know the square feet, look for the number of bedrooms. If the 1,200-square-foot house with the orchard has three bedrooms, the living spaces left might make for a snug fit. If it has only one bedroom, however, the place might well feel palatial!

EARTHQUAKES

The first number that any news service picks up when reporting an earthquake is its magnitude on the Richter scale, a 1 to 8 scale that is extremely difficult to translate into actual effects. How great *is* a 3.5 earthquake, and is a 7.0 earthquake twice as great, or more? Most of us would like to know at what point life and/or property are threatened.

The first scale used to measure earthquakes did just that. It was devised by Italian seismologist Giuseppe Mercalli in 1902 and revised by American scientists in 1936. Mercalli interviewed earthquake survivors, and then used a 1 to 12 (I to XII) scale to describe the effects. It was much like Beaufort's scale describing the effects of wind speeds (see **Wind,** page 206). Here it is, with approximate Richter figures added:

THE MERCALLI SCALE

Mercalli	Mercalli Characteristics	Approx. Richter
I	Detectable only by seismographs	Less than 3.5
II	Feeble: noticed only by some people at rest	3.5
III	Slight: similar to vibrations of a passing truck	4.2
IV	Moderate: felt indoors; parked cars rock	4.5
V	Rather strong: felt generally; sleepers wake	4.8
VI	Strong: trees sway; furniture moves; some damage	5.4
VII	Very strong: general alarm; walls crack	6.1
VIII	Destructive: weak structures damaged; walls fall	6.5
IX	Ruinous: some houses collapse as ground cracks	6.9
X	Disastrous: many buildings destroyed; rails bend	7.3
XI	Very disastrous: few buildings survive; landslides	8.1
XII	Catastrophic: total destruction; ground forms waves	Greater than 8.1

Although Mercalli's scale is still useful for getting a general idea of earthquake effects, it lacks accuracy—damage done by earthquakes is influenced by many factors, including population density, type of ground involved, building structures, etc. Furthermore, it does not measure the earthquake itself. Seismologist Charles F.

Richter was bothered by all this and devised a system, which he never intended to become so universally used, to measure the actual "magnitude" of the quake, a term he used to take the emphasis off the effects (intensity) and place it on the physical activity—the earth's action—during the quake.

The result was the Richter scale, sometimes misleading because it is not actually a 1 to 8, but a 1 to 10 million scale. Each level indicates a tenfold increase in magnitude from the level before, making a 6.0 quake not twice as great as a 3.0 quake, but a thousand times greater, and an 8.0 *ten million* times greater than a 1.0. Below is a translation of the Richter scale:

THE RICHTER SCALE

Richter Number	Increase in Magnitude
8	10,000,000
7	1,000,000
6	100,000
5	10,000
4	1,000
3	100
2	10
1	1

Richter measured earthquakes using a seismograph to record the size and time of tremors. He measured an earthquake by relating the *amplitude* or size of the tremors (measured in millimeters on seismograph records) to the time between the primary (first) and secondary (second) tremors, much like an obstetrician judging the stage of labor by timing contractions and analyzing their intensity.

Earthquakes are scary, and no country takes them more seriously than the United States. The world's foremost collection of earthquake data lies in Golden, Colorado, at the U.S. Geological Survey's National Earthquake Information Center. Here 60,000 seismic readings are collected every month from 12 seismograph installations across the United States, 650 stations around the world, and many more which give information when called upon. Most rescue operations to any earthquake-hit area of the world begin with a signal from Golden.

Unfortunately, even sophisticated electronic ears listening for ground activity still can't *predict* earthquakes.

ELECTRICITY

It's time to sort out those one-syllable words that have been confounding you all these years—words like *amps, volts, watts,* and *ohms* that are often preceded by a number probably meaningless to you (unless it's translated into dollars on your power bill). If you are like many people, the best you've done so far is figure out that a 100-watt light bulb is brighter than a 40-watt light bulb (even if it often costs the same).

To begin with, *amps,* short for amperes, are the measure of actual electrical current (electrons per second) that flows along the wires in your house.

Volts are the measure of the pressure pumping the amps into, say, your house. (If amps were water, volts would be the water pressure forcing the water through the pipes.) High voltage pumps move amps along a wire more rapidly than low voltage. Normal voltage for most homes is 120 volts.

Ohms is a measure of the resistance that slows the flow of amps. The higher the ohms—the more resistance created—the fewer amps can get past. Ohms (or resistance) can be increased by using, among other objects, thinner wires or "resistors," devices that control the amount of current that can enter an electrical appliance. These are very important, as few appliances require the full force of 120 volts.

Watts are a measure of work—the amount of work an appliance is capable of. If you continue with the water analogy, watts are the power that amps (water) have when propelled at certain volts (pressure) past a resistance of a certain number of ohms (through a hose of a certain size and coming out of a nozzle with a particular opening). So the resistance in a 40-watt light bulb lets in less current than the resistance in a 100-watt light bulb, producing fewer watts.

This can all be summed up in a few little equations:

$$\text{watts (power)} = \text{volts (pressure)} \times \text{amps (current)}$$

OR

$$\text{amps (current)} = \frac{\text{volts (pressure)}}{\text{ohms (resistance)}}$$

OR

1 watt = 1 amp flowing at 1 volt pressure

OR

1 volt will drive 1 amp through the resistance of 1 ohm.

You pay for electricity by the *kilowatt hour,* which is the amount of energy required to produce 1,000 watts of power (*kilo* means *thousand*) for 1 hour. Check your electric meter: some meters simply show a number, while others show four or five little "clocks," which are read from left to right. A meter reader comes around once a month or so to record this number. Last month's reading is subtracted from the current reading so the electric company will know how much you've used.

HOW TO READ YOUR ELECTRIC METER

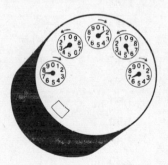

Because the dials are geared to each other, the pointers on some dials turn clockwise; on others, counterclockwise. Read the dials from left to right. When a pointer is between two numbers, always read the smaller number. The above reading is 7-3-1-5-6 kilowatt-hours.

Source: Pacific Gas and Electric Company. Used with permission.

How much does it actually cost you to run those indispensable appliances in your house? First, check your electric bill for the amount you are charged per kilowatt hour—10 cents may soon be common. Then find out how long an appliance takes to use up a kilowatt hour (1,000 watts/hour) by dividing 1,000 by the wattage of the appliance. For example, a 200-watt television will give you 5 hours of entertainment for 1 kilowatt hour of electrical power (1,000 divided by 200). Divide the cost per kilowatt hour (10 cents) by the number of hours the TV runs for that amount (5 hours), and you can determine your cost per hour—in this case 2 cents. It's surprising how much—or how little—power appliances really use.

ESTIMATED KILOWATT APPETITES OF SOME COMMON APPLIANCES

Appliance	Kilowatts per hour
Light bulbs	0.001 per watt
Electric blanket	0.07
Stereo	0.10
Television (color)	0.23
Clothes washer	0.25
Evaporative cooler	0.50
Forced-air furnace fan	0.50
Coffee maker	0.60
Vacuum cleaner	0.75
Iron	1.00
Dishwasher	1.00
Toaster	1.20
Oven	1.30
Air conditioner (1,500 watts)	1.50
Hair dryer (blower)	1.50
Microwave oven	1.50
Range-top burner	2.80
Waterbed (with thermostat)	2.80
Waterbed (no thermostat)	3.90
Clothes dryer	4.00
Refrigerator/freezer (16 cubic feet)	5.00
Frostless freezer	5.10
Freezer (manual defrost)	3.00
Refrigerator/freezer (22 cubic feet, side by side)	7.00

Appliance	Kilowatts per hour
Water heater (electric)	16.00

Source: Pacific Gas & Electric Company. Used with permission.

ENGINES (HORSEPOWER)

Horsepower is one of those charming measures that drive scientists crazy for it is based on an arbitrary, though lively, standard: the strength of a particular horse. One day Scottish inventor James Watt rigged up the horse of his choice with a rope and some pulleys so he could demonstrate the power (work capability) of his recently invented steam engine (patented in 1783). When Watt fastened a weight to one end of the rope and the horse to the other, he found that the horse, by moving forward, could raise a 3,300-pound weight 10 feet into the air in one minute.

If Watt had increased the weight ten times to 33,000 pounds, the horse would have raised it only one foot in one minute. Watt called the amount of work *1 horsepower*: one horsepower was equal to 33,000 foot-pounds per minute, a measure that is still used today, although more commonly referred to as 550 foot-pounds per second (33,000 divided by 60 seconds). As horses were one of the main energy sources of that day, these were terms the general public could understand.

Watt labeled his steam engines in equivalent horsepower—a 10 horsepower engine could do the work of 10 horses; i.e., could lift 5,500 pounds per second. Today the work capability of many engines is still labeled in horsepower. The horsepower of an engine results from a variety of factors, including the size of the engine, measured in cubic centimeters (cc's—see **Motorcycles,** page 124), engine design, etc. Note that if the horse can't lift the weight, no work is done and no power is developed. The horse provides the force: work = force × distance; power = force × distance/time. So you might say that a 100 hp engine can work 100 times faster than a horse.

It is interesting to think in terms of actual horses doing the work of the following common gasoline engines:

Garbage disposal	½ horsepower (hp)
Power lawn mower	3 hp
Riding lawn mower	8 hp
Motorcycles	10–100 hp
Small one-engine airplane	30 hp
Small car	75–80 hp
Passenger car	80–300 hp
Small trucks	100 hp and up
Sports cars	as much as 200 hp
Small 6-passenger 2-engine airplane	400 hp/engine
B36 4-engine bomber	3,650 hp/engine

Horsepower can, by the way, be translated into "watts," another way to measure power, not invented by Watt but named in his honor anyway (see **Electricity,** page 60): 1 horsepower = 746 watts.

EXPONENTS

If your mind shuts down when you see a number like 5.88×10^{12} (the number of miles light travels in one year), fearing some form of higher mathematics, you're in for a surprise. Exponents, as those small, elevated numbers are called, simply tell you how many times to multiply the number it modifies by that number (e.g., $4^3 = 4 \times 4 \times 4 = 64$). Exponents of ten work the same way and make it easier to count the zeros in tediously large or small numbers. The above number written out is awkward: 5,880,000,000,000 miles. Even worse, a beta ray particle has a mass of 0.00000000000000000000000000091 gram, more neatly written as 9.1×10^{-28} gram.

Here's how to read exponents of ten. For positive exponents, simply use as many zeros as the exponent reads:

$10 = 10^1$	(ten)
$100 = 10^2$	(one hundred)
$1,000 = 10^3$	(one thousand)
$10,000 = 10^4$	(ten thousand)
$100,000 = 10^5$	(one hundred thousand)
$1,000,000 = 10^6$	(one million)
$1,000,000,000 = 10^9$	(one billion)
$1,000,000,000,000 = 10^{12}$	(one trillion)
and so on ...	

For negative exponents, add the zeros numbered in the exponent on the left, *less one,* or, more simply, move the decimal as many places to the left as numbered in the exponent. So the negatives look like this:

$0.01 = 10^{-2}$	(one hundredth)
$0.001 = 10^{-3}$	(one thousandth)
$0.0001 = 10^{-4}$	(one ten-thousandth)
$0.00001 = 10^{-5}$	(one one-hundred-thousandth)
$0.000001 = 10^{-6}$	(one millionth)
$0.000000001 - 10^{-9}$	(one billionth)
$0.000000000001 = 10^{-12}$	(one trillionth)
and so on ...	

Most numbers involving exponents are written as a multiplication problem—the second number simply tells you how many places to move the decimal in the first number. When the exponent is positive, move the decimal point in the first number that many places to the right; e.g., in 5.6×10^6, move the decimal six times to the right, making 5,600,000. When the exponent is negative, move the decimal point that many places to the left, e.g., in 5.6×10^{-6}, move the decimal in 5.6 six places to the left, making 0.0000056.

Now try these:

A jumbo jet weighs about 3.75×10^5 kilograms.
A house spider weighs 10^{-4} kilogram.
Neptune is 2.677×10^{12} miles from the earth.
The mass of a hydrogen atom is 1.66×10^{-24} gram.

FABRIC CARE

You may not want to know that there are seventeen commonly used ways to clean fabric. But if you're wondering how to care for that special garment or the length of fabric you painstakingly sewed up yourself, the letter or number inside a triangle printed on the end of the fabric bolt or sometimes on the labels of ready-made clothing, may save you grief. The number or letter code stands for a set of washing or cleaning directions which are not always included with the code. A glance at the list below will tell you to beware of fabrics with a letter code—they usually require special attention that you might not always be in the mood to give.

Although the law no longer requires the use of these fabric care codes—the care instructions themselves are now supposed to be printed on fabric boards—the codes are still in use.

CARE LABELING INSTRUCTIONS

Code	How to Clean
Method 1	Machine wash, warm.
Method 2	Machine wash, warm; line dry.
Method 3	Machine wash, warm; tumble dry; remove promptly.
Method 4	Machine wash, warm, delicate cycle; tumble dry, low; use cool iron.
Method 5	Machine wash, warm; do not dry-clean.
Method 6	Hand wash separately; use cool iron.
Method 7	Dry-clean only.
Method 8	Dry-clean, using special method for pile fabrics.
Method 9	Wipe with damp cloth only.
Method B	Machine wash, warm, separately. Do not use light-colored pocketing or waist-banding, remove white or light trim.

Method C	Machine wash, warm; tumble dry. Use detergent only; do not use soap.
Method E	Machine wash, cold water, separately; do not use light-colored pocketing or waist-banding; remove white or light trim.
Method G	The inherent characteristics of this fabric will cause it to fade after each washing. The color will also tend to rub off. Must be washed before wearing. Wash and dry separately. Do not dry-clean.
Method H	Hand wash, cold water only.
Method J	Machine wash, warm. Tumble dry. Home launder only, preferably in soft water. Use phosphate detergent powder or nonphosphate liquid detergent. Do not use soap, nonphosphate powdered detergents, bleach, or fabric softeners.
Method K	Dry-clean only, machine temperature not to exceed 100 degrees F. Do not use coin-operated unit. May be wiped clean with a damp cloth; for more stubborn stains, wipe with cloth using a mild detergent and water. Do not press. If touch-up ironing is required, press back of fabric with a warm (not hot) dry iron.
Method L	Dry-clean only, machine temperature not to exceed 100 degrees F. Do not use coin-operated unit. Spot cleaning not recommended. If required, sponge or blot; do not rub. If touch-up ironing is required, press back of fabric with a warm (not hot) dry iron.

*F*ABRIC *WIDTHS*

Unless you're in the textile or clothing industry, what you think of as "fabric" is sold by the yard or fraction thereof—less often by the inch—from "bolts" at the fabric store, no matter what kind of fabric you're talking about, how much you want, or what the width is. It's not really the length of fabric that confuses people, however; it's the width. The most common width in the United States is 44 to 45 inches, which probably comes from the English *ell,* a 45-inch measure used for years to measure fabric. (The ell was also used for other purposes and varied in length from country to country, purpose to purpose.) England and the United States are about the only countries making fabric in 44- to 45-inch widths. The trend, however, is toward 60-inch fabric, which is good news because it can be used more efficiently than narrower widths—most patterns can be cut from 60-inch fabric with less waste, using less yardage.

You'd think that something like fabric would be sold by the square yard, like carpeting, or that there would at least be standard widths. Many fabrics, however, are offered in 41- or 54-inch widths, and it's not unusual to find imported fabrics in 36-inch widths, or lingerie tricots up to 108 inches wide. The problem comes in trying to decide how many yards you need of a particular fabric. Many patterns estimate the yardage needed of more common widths of fabric. This information is more accurate these days than it used to be, when you could almost always do with less fabric than the pattern called for. You don't have to guess now—computers do the estimating and they are quite accurate. If, however, you're considering a less common width—say, a 36-inch—you may need some help.

The Yardage Conversion Chart here is offered in the spirit of the true emergency—when there is absolutely no way you can lay your pattern out on the fabric (that is the only accurate way to know how much fabric you'll really need). It's also true that some pat-

YARDAGE CONVERSION CHART

Width	32"	35"–36"	41"	44"–45"	52"–54"	58"–60"
Length in Yards	1⅞	1¾	1½	1⅜	1⅛	1
	2¼	2	1¾	1⅝	1⅜	1¼
	2½	2¼	2	1¾	1⅝	1 ³⁄₈
	2¾	2½	2¼	2⅛	1¾	1⅝
	3⅛	2⅞	2½	2¼	2¼	1¾
	3⅜	3⅛	2¾	2½	2¼	1⅞
	3¾	3⅜	2⅞	2¾	2¼	2
	4	3¾	3⅛	2⅞	2⅜	2¼
	4⅜	4¼	3⅜	3⅛	2⅝	2⅜
	4⅝	4½	3⅝	3⅜	2¾	2⅝
	5	4¾	4	3⅝	2⅞	2¾
	5¼	5	4⅛	3⅞	3⅛	2⅞

Developed by Rutgers Cooperative Extension Service, Rutgers, The State University of New Jersey.

terns cannot be cut from narrower fabric than called for; or you may need just as much of a wide width as you would of a narrow one. To compare yardage of different widths of fabric, read along the same horizontal line. For example, a pattern requiring 1⅜ yards of 54-inch fabric will probably need 2 yards of 36-inch fabric. (See second line of chart.)

FERTILIZERS

Looking at the fertilizer shelves at the local plant store can make a person want to jam his green thumb in his pocket, find his car keys, and leave. The array of offerings is staggering. How can one possibly decide which one to buy?

It helps a little to understand the numbers. There are three numbers on every bag, box, and bottle of plant food (fertilizer), usually found on the label in large print, separated by hyphens; e.g., 10-10-5. These numbers represent the percentage, by weight, of the three chemical elements essential to plant growth contained in the fertilizer: nitrogen, phosphorus, potassium, always in that order. This means that the above 10-10-5 fertilizer contains 10 percent nitrogen (N), 10 percent phosphorus (P), and 5 percent potassium (K). What's left—the other 75 percent, in this case—is inert (nonactive) matter.

What's important here is not how high the numbers are but the *ratio,* or balance, of the three numbers. For example, a 5-5-5 fertilizer is essentially the same as a 10-10-10 or a 30-30-30 fertilizer: they all have a 1-1-1 balance, and they will all have essentially the same effect. The difference between a 10-10-10 fertilizer and a 30-30-30 fertilizer is (1) the amount and/or frequency of applications, and (2) the price. The ratio you choose in a fertilizer depends on what you're feeding, your soil, and what kind of growth you may be trying to encourage: The first number (nitrogen) affects stem and leaf growth, the second number (phosphorus) encourages root growth, and the third number (potassium) is needed for flowers, fruits, and general all-around sturdiness.

In truth, there's so much disagreement on which combination of nutrients is best for what that even a fairly knowledgeable gardener may be tempted to toss in the trowel trying to decide what to use. For example, a recently sampled fertilizer shelf offered three different plant foods for tomatoes. Here's the breakdown:

THREE TOMATO PLANT FOODS

Fertilizer	Amount	Cost	N-P-K	Amount needed	Frequency
Brand 1	12 spikes	$1.89	8-24-8	2 per plant	every 2 months
Brand 2	1½ lb.	$4.69	18-18-21	1 tbsp./gallon water	every 7–14 days
Brand 3	4 lb.	$2.15	8-12-6	⅞ cup/plant ⅛ cup/plant	at planting every 6 weeks after

The disparity of price, pounds, and nutrients here is boggling! The slow-release spikes will cost you 63 cents per plant (4 per plant per season) which seems quite high. But try to compare that cost to the others—you need to know how many tablespoons are in a pound (Brand 2), and how many cups are in a pound (Brand 3), a figure that will differ from plant food to plant food. Then how do you decide on the nutrient balance? You're tempted to go with the highest numbers until you notice that you use only a tablespoon of it, but you use nearly a cup (at planting) of the one with the lowest numbers. Which ends up feeding the plant the most?

It's important to remember that it's the balance that's important, not high numbers. For tomatoes, you may decide that a fertilizer high in potassium (good for fruiting and flowering) may be a good choice, but then you may wonder why the others, also made specifically for tomatoes, offer a different balance—potassium is actually the lowest number in Brand 3!

If you can't figure out what you want, pick a general-purpose fertilizer that you can dump on everything without thinking about it. The recommendations for this differ widely, of course—some experts recommend a 1-2-2 ratio (such as 5-10-10), others a 1-1-1 ratio (such as 10-10-10), and an expert from the National Fertilizer Development Center suggests that a 4-1-2 ratio (such as 20-4-10) would suit most lawns and garden plants. Your soil, of course, may determine which to use, so you may want to have it tested if you're really serious about all this. It may be comforting to know, though, that an overdose of phosphorus (P) or potassium (K) is not harmful to most plants. (An exception is zoysiagrass, which doesn't like too much P.) Other suggestions include asking the person who has the best garden in the neighborhood (your soils may be similar) what he or she uses, calling up your county cooperative/extension agent (see **Soil, Garden,** page 166), who may offer comforting advice, and/or starting a compost heap. (See pages 168–69 for a list of cooperative extension offices.)

You should know that there are three kinds of fertilizer: organic, inorganic, and slow-release fertilizers. Organic fertilizers are slow-acting but long-lasting; most chemical fertilizers are fast-acting but require more frequent applications, with the exception of slow-release fertilizers. Slow-release fertilizers are useful for shrubs, trees, and lawns. Urea formaldehyde (sometimes called ureaform or nitraform) is a synthetic organic fertilizer (which sounds like a contradiction in terms). It dissolves more slowly than other fertilizers; it really is slow, sometimes taking two years to become available to plants. A sulfur-coated urea, developed by the Tennessee Valley Authority, is one of the cheapest slow-release fertilizers and an effective once-a-year nitrogen feeding for lawns.

Here are the ratios of some of those odd-sounding offerings at the plant store:

SOME N-P-K FERTILIZER RATIOS

Ammonium nitrate	33-0-0
Ammonium sulfate	21-0-0
Blood meal	13-0-0
Bone meal	1-23-0
Fish emulsion	5-1-1
Horse manure	1-1-1
Muriate of potash	0-0-60
Sulfate of potash	0-0-52
Super phosphate	0-20-0
Urea	46-0-0

Now, if you're still looking at the shelf, here are a few of the many recommendations from a wide variety of master gardeners and plant food companies. This list will give you an idea of what sort of balance different plants may require, although your soil will influence what you need in a fertilizer. To use a plant food (when and how much), follow the directions on the package or that of your chosen expert.

OTHER NITROGEN-PHOSPHORUS-POTASSIUM RECOMMENDATIONS

The most important consideration here is the *balance* of nutrients, not how high the numbers are, so only the lowest commonly used numbers of that balance will be mentioned. In other words, 15-30-15 fertilizer will be listed as 5-10-5, since that is the essential balance; higher numbers that provide that balance will work as well.

African violets	10-10-5 or 12-36-14
Azaleas, camellias, rhododendrons	10-8-7
Bulbs	4-12-8
Cacti and other succulents	5-10-5
Citrus	10-12-4 or 6-9-6
Evergreens	14-7-7 or 10-10-10 or 16-4-4
Ferns	fish emulsion (5-1-1)
Flower gardens	5-10-5 or 5-10-10 or 12-55-6
Houseplants (foliage)	5-10-5 or 12-6-6
Houseplants (flowering)	5-10-5
Lawns	5-10-5 or 10-6-4 or 36-6-6 or 22-3-3 or 8-12-24, etc., etc., etc.
Perennials	0-20-20
Roses	5-10-5 or 8-12-4
Trees and shrubs	10-6-4 or 12-4-8 or 20-10-5
Vegetables (good general)	5-10-5 or 10-10-10
Beans, beets, broccoli, chard, cucumbers, eggplant, melon, onions, peppers, potatoes, spinach, tomatoes, turnips	5-10-5
Brussels sprouts, carrots, peas, sweet potatoes	5-10-10
Leafy vegetables, cauliflower, celery, corn, radishes	10-10-10
Vines	5-10-5 or 0-10-10 (for less foliage)

FINANCIAL INDEXES

Newspapers and magazines frequently report statistics collected by agencies and departments of the federal government, reported by using an index system that looks odd if you don't know how to read it. Such statistics are not hard to read, though, and are worth the few minutes it takes to figure out—many of these numbers are of great interest to the nation's financiers as well as to those of us interested in the economy. Examples like the Consumer Price Index and the Gross National Product are explained in separate entries, but there are many others, like the Economic Forecast monthly report (from the U.S. Department of Commerce) and Industrial Production report (from the Federal Reserve Board) shown below. Examine these two:

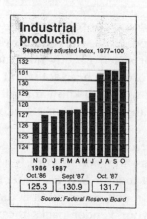

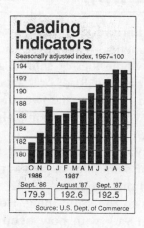

These are typical newspaper presentations of the figures. All you need to remember is that all these indexes compare the latest figures to those of an earlier time, or a series of earlier times, in order to show growth or lack of it. The year the figures are being compared to is at the top of each index: Leading Indicators uses 1967 as its base, while Industrial Production compares its figures to 1977's. Any number over 100 shows percentage of growth; numbers under 100 show percentage of decline. The letters at the bottom of the graph represent the previous twelve months, and the index is given for each. For example, the September 1987 Economic Forecast reading 192.5 means that the four items used to make up this index (including stock prices and other factors) were 92.5 percent more active than in 1967. The Industrial Production Index shows that in October 1987 industrial production was up 31.7 percent over that of 1977.

Indexes of this type are not difficult to read once you learn to subtract 100 from the figures to find the percentage of growth. Compare this to the base year or to more recent periods, even the previous month, to get an idea of where the economy might be heading.

FIREWOOD

Ordering firewood is rarely a welcome chore—there are just too many anxieties: Will you be cheated? How will you know? Should you get whole logs or split? Big or small? Exactly how much is a "cord" or a "truckload" or a "face cord"? Is cheap softwood a better deal than expensive hardwood? How do you know if a full cord of pine at $75 is more economical than a face cord of 18-inch oak for the same price?

Firewood is usually sold by the *cord,* an inexact measure but useful for something as awkward as firewood. A cord is a stack of firewood that measures 8 × 4 × 4 feet, or 8 × 8 × 2 feet, which amounts to the same thing. How much firewood is actually in a cord can vary with how tightly your vendor packs the wood— although a cord contains 128 cubic feet, the actual wood in it may measure anywhere from 65 to 100 cubic feet, depending on the air spaces. Some people are convinced that small logs make a tighter cord, others argue that large logs make a heavier cord, but most people agree that big logs with small logs filling in the spaces between them is best. (Old-time wisdom has it that the holes can be big enough to let a squirrel through, but not a cat.)

Many reputable firewood dealers have their firewood stacked on their lots in measurable fashion where you can examine it; others have figured out what fraction of a cord fits in their delivery truck. How do you know how much you are getting? You may have to take the dealer's word for it, a situation that does little to inspire trust. You can restack the firewood and measure it. If the dimensions seem disturbingly off, call this to your vendor's attention.

Another confusing but popular measure of firewood is the *face cord,* also known as the *short cord.* The face cord is a stack of logs measuring 8 × 4 feet × whatever the length of the logs. So if you have 2-foot logs, you'll have half a cord (8 × 4 × 2 feet), but 16 or 25 inches or any other length log still makes a face cord if the "face," or outside, measures 4 × 8 feet. How do you compare the price of a face cord to that of a full cord? Look up the cubic feet for varying face cords below—or figure it yourself by finding the cubic inches (96 × 48 inches × the length of the logs in inches) and dividing by 1,728 (the number of cubic inches in a cubic foot)— and divide this figure into the price. For example, a face cord of 14-inch logs is 37.3 cubic feet. Divide the cubic feet into the price— say, $40—and you find you are paying $1.07 per cubic foot. Compare this to a full cord for, say, $105; $105 divided by 128 cubic feet (one cord = 128 cubic feet) gets you $0.82 per cubic foot, a much better deal.

CUBIC FEET FOR SOME FIREWOOD MEASURES

Full cord	128 cu. ft.
Face cord of 12″ logs (¼ cord)	32 cu. ft.
Face cord of 14″ logs	37.3 cu. ft.
Face cord of 15″ logs	40 cu. ft.
Face cord of 16″ logs	42.6 cu. ft.
Face cord of 18″ logs	48 cu. ft.
Face cord of 20″ logs	53.3 cu. ft.
Face cord of 22″ logs	58.6 cu. ft.
Face cord of 24″ logs (½ cord)	64 cu. ft.
Face cord of 25″ logs	66.6 cu. ft.
Face cord of 30″ logs	80 cu. ft.

Firewood sold by the fraction of a cord—two-thirds is typical— can also be figured in cubic feet. However, some people claim that short logs pack tighter than long logs, so that a cord of 24″ logs contains more wood by weight than a cord of 48″ logs. This is typical of the whole firewood business—there is a whole lot of opinion, and in the end you just have to trust your own instincts, especially if you run into some of the other firewood measures that pop up.

Although in some states it's illegal to sell firewood by anything but the cord or fraction thereof, one still sees it for sale by the "truckload," sometimes called a *run*, which is as much as that size truck can carry. Although the rule of thumb is that a ½-ton pickup can carry ½ cord and a 1-ton pickup a full cord, it really depends on the weight of the wood, the height of the truck bed's sides, and whether special frames have been mounted. If you are truly un- lucky, you may also be faced with "stove cords," "fitted cords," or "fireplace cords." With any of these "measures," the only way to know for sure how much you're getting is to know you can trust your dealer or to stack the stuff yourself and figure the cubic feet. One clue may be the weight: A typical cord of hardwood weighs 1½ to 2 tons. Anyone who claims to dump a cord of hardwood on your property with one ½-ton truckload is fudging. (See pages 77– 78 for weights of some other woods.)

In the end, the best measure of firewood may be the one most rarely used—*weight.* The amount of heat you get from your fire- wood is the same *per pound* for *all* firewood—one source says you will have about 7,000 Btu's (see page 84) from a pound of wood, while another source puts the figure at 6,400 Btu's. The important thing is this: No matter what kind of wood you use, you usually get the same heat value *per pound.* The denser—or harder—the wood, the heavier it is and the more pounds—and heat—you'll have per cord. The only true way to evaluate your wood is by price per pound, which you can figure out if you know the average weight per cubic foot of your firewood. *Divide the price* per cubic foot of firewood *by the pounds* per cubic foot that it weighs, and then you will know what you are paying for every 6,400 Btu's of heat. For example, a $100 cord of white fir costs 78 cents per cubic foot ($100 divided by 128 cubic feet). White fir weighs 28 pounds per cubic foot, so one pound costs about 3 cents, producing 6,400 Btu's of heat. This is just an educated guess—the cubic feet you figured above is not solid wood but stacked wood with air holes, the firewood must be dry (figures below are based on a 20 percent moisture content), and the cords must be honest—but it is an interesting way to compare firewood deals or the value of one kind of wood with another.

APPROXIMATE WEIGHT PER CUBIC FOOT OF SOME COMMON WOODS

Types of Hardwood	Lbs./cu.ft.	Types of Softwood	Lbs./cu.ft.
Alder, red	38	Bald cypress	32.5
Apple	48.7	Cedar:	
Ash:		Alaska	32.5
Black	46	Atlantic-white	24
Green	41.5	Eastern red	34
White	43	Northern white	23
Aspen, quaking	27	Port-Orford	30.5
Basswood	25	Western red	24
Beech	44	Douglas fir	33–35
Birch:		Fir:	
Paper	37.5	Balsam	26
Sweet	46	California red	27.5
Yellow	43	Grand	27
Butternut	27.5	Pacific Silver	31
Cherry	36.5	Subalpine	24
Chestnut	31	White	28
Cottonwood	24	Hemlock:	
Elm:		Eastern	29
American	36	Western	32.5
Rock	45	Mountain	32.6
Slippery	37.5	Larch	37.5
Hackberry	38	Pine:	
Hickory, pecan	47	Eastern white	26
Hickory, true:		Jack	31
Mockernut	51	Loblolly	36.5
Pignut	53	Lodgepole	36.5
Shagbark	51	Longleaf	42
Shellbark	50	Pitch	36.5
Honey locust	47	Pond	40
Locust, black	53	Ponderosa	29
Magnolia, southern	36	Red	32
Maple:		Sand	36
Bigleaf	34	Shortleaf	36.5
Black	41	Slash	42
Red	38	Spruce	32
Silver	34	Sugar	26
Sugar	44	Virginia	35
Oak, red:		Western white	27
Black	44	Redwood:	
Cherrybark	48	Old-growth	29
Laurel	44	Young-growth	26
Northern red	44	Spruce:	
Pin	43	Black	29

Types of Hardwood	Lbs./cu.ft.	Types of Softwood	Lbs./cu.ft.
Oak, red:		Spruce:	
Scarlet	47	Engelmann	25.5
Southern red	41	Red	29
Water or Willow	44	Sitka	28
Oak, white:		White	28
Bur	45.5	Tamarack	38
Chestnut	45		
Live	67		
Post	47		
Swamp chestnut	47		
Swampy white	51		
White	47		
Sassafras	32.5		
Sweet gum	36		
Sycamore	36		
Tan oak	45.5		
Tupelo	36		
Walnut, black	40		
Willow, black	27.5		

Adapted from specific gravity figures for hardwoods and softwoods offered in Agriculture Handbook No. 72, Wood Handbook: Wood as an Engineering Material, by Forest Products Laboratory, Forest Service, U.S. Department of Agriculture, 1974.

Of course, many woods not found in that list are excellent for firewood. For example, in California, madrone and eucalyptus are popular, being heavier than oak. Fruitwoods, such as apple, are also oak equivalents. As for oak, there are more than one hundred kinds in one California county alone. On the whole, hardwood is usually preferred to softwood for hotter fires and more efficient burning.

FOOD (ENERGY VALUE)

You may think that your food contains calories, or perhaps is even made of calories, but this is not true. The calorie count isn't a measure of the content of food, but of the energy value of food. It's confusing, because not only are calories metric, but there are calories and there are Calories, and they are not equal.

A calorie is a measure of energy, and it comes in two sizes, small and large. The small calorie, used by scientists and by metric-system countries to measure heat/energy, is defined as the amount of energy needed to raise the temperature of 1 gram of water 1 degree centigrade, or from 14.4 degrees C to 15.5 degrees C. It takes 1,000 of these small calories to make a kilocalorie, also known as the large Calorie—or the amount of energy it takes to raise 1 *kilo*gram of water 1 degree centigrade. It is this large Calorie that is used to measure the fuel value of food; i.e., the amount of energy it takes to raise 4 pounds of water 1 degree Fahrenheit. Since the small calorie isn't commonly used in this country, we don't bother to capitalize the "c."

Food is potent. A pound of TNT, for example, has the same energy value as your average fast-food quarter-pound cheeseburger (500 calories). It's quite possible, however, that you think of calories not in terms of energy, but in terms of fat, and you're not far off. Each pound of you represents about 3,500 calories. To loose a pound, you need to take in 3,500 fewer calories than you burn. But before you start a diet, take a look at where the calories are that you're eating—a little rearrangement can cut out quite a few without reducing your meal size much. Naturally, you should never go on a diet without first checking with your doctor.

You consume calories from the three basic types of foods you eat: fats, proteins, and carbohydrates. A gram of protein has an energy value of 4 calories, and a gram of carbohydrate has an energy value of 4 calories, but a gram of fat has 9 calories, more than twice that of the first two. This is worse news than you think—it has been estimated that 40 percent of the calories in the typical American diet is fat. If you are typical, another 20 percent of your daily calories are simple carbohydrates (sugars), leaving only 40 percent of your diet for protein and complex carbohydrates (grains, fruits, and vegetables). Here's the recommended balance:

RECOMMENDED CALORIE BALANCE

12%	protein
20–30%	fat
53–63%	complex carbohydrates (grains, fruits, vegetables)
5–10%	simple carbohydrates (sugar, syrup)

One of the simplest ways to cut calories is to eat less sugar and fat, remembering that fat, including butter and margarine, adds up quickly—about 100 calories per tablespoon—and that sugar contains 48 calories per tablespoon. You can probably cut more than 300 calories from your daily intake by eliminating most of your sugar and fat, without affecting the size of your meals.

One of the most effective ways to cut calories, according to many experts, is to burn them off with exercise. In fact, studies have shown that if you don't exercise when you diet, your body may adjust its metabolism to fewer calories. This means that if you normally eat 2,200 calories but cut that to 1,200, your body may adjust so completely that, although you lose weight at first, after a while 1,200 may become your weight maintenance amount and any calories you take in over that could put weight on you. Ironi-

CALORIES TO BURN

Activity	Calories burned per half hour	Activity	Calories burned per half hour
Sleeping	30–36	Walking 3.5–4 mph	150–222
Watching TV		Weeding	
		Scrubbing floors	
		Bicycling	
Doing sedentary work/ play	Up to 75	Skiing	
Driving the car		Playing tennis	
Playing a musical instrument		Dancing	
Sewing			
Ironing			
Level walking, 2.5–3 mph	75–147	Backpacking uphill	223–360
Fixing the car		Playing basketball	
Doing carpentry work		Playing football	
Shopping with light load		Swimming	
Playing golf		Climbing	
Sailing			
Playing Ping-Pong			
Playing volleyball			

Source: Recommended Dietary Allowances. *Used with permission from the National Academy Press, Washington, D.C.*

cally you often end up heavier than you were before you began the diet. The effective way to lose weight, many experts say, is to combine a smaller calorie cutback from the right places with daily or frequent exercise. And that's good news, because exercise has great side effects, like keeping your bones and heart strong, your cholesterol down, and your spirits up.

This is not a diet book, but try a little math: If you burn off an extra 200 calories per day with exercise, and cut out 300 calories, that's 500 per day, enough to let you lose a pound a week! This is 10 pounds in 10 weeks, without cutting your meal size by much or by going on a crash diet that may make you bigger than when you began.

FOOD GRADING

A TV quiz show host recently asked whether "Choice" or "Prime" was the better beef, and the contestant didn't know. Not surprising, since the U.S. Department of Agriculture, responsible for most food grading, uses one system for beef and lamb, another for pork, and yet another for poultry. In addition, fresh fruits and vegetables are graded differently from canned ones, and eggs have yet a different grading—buy Grade A poultry and you get the best, but Grade A eggs are only second best.

TYPICAL FEDERAL GRADES FOR FOODS

Food	Best	2nd Best	3rd Best
Beef, lamb, and veal	Prime	Choice	Good
Pork	Acceptable	Unacceptable	—
Poultry	A	B	—
Fish*	A	B	C
Eggs and butter, Cheddar cheese	AA	A	B
Fresh fruits & vegetables**	U.S. Fancy	U.S. No. 1	U.S. No. 2
Canned fruits & vegetables, frozen juices, jams, and jellies**	A	B	C

* A voluntary inspection program by U.S. Department of Commerce.
**These designations may vary for some products.

Source: Agriculture Marketing Service, U.S. Department of Agriculture.

Inspection stamps seem to be a waste of time unless consumers have a clear idea of what they mean. Your only recourse may be to check the chart on page 81.

How do you know the grade on food products, especially fresh produce? Most of the time, you don't; however, many quality stores will tell you if you ask—the grade is usually printed on the cartons produce is shipped in.

COMMON GRADE MARKS

Meat
Beef. Lamb. Veal.

Poultry
Chickens. Turkeys. Ducks. Geese.

Fruits and Vegetables and Related Products
Fresh fruits and vegetables.

U.S. Fancy (This grade name is more likely to be found without the shield.)

Canned and frozen fruits and vegetables. Dried or dehydrated fruits. Fruits and vegetable juices, canned and frozen. Jams, jellies, preserves. Peanut butter. Honey. Catsup, tomato paste.

Eggs

Source: Agricultural Marketing Service, U.S. Department of Agriculture.

GAS

Every home using natural gas has a gas meter, called a diaphragm type displacement meter, which measures in cubic feet how much gas you use. It does this by moving the gas through several chambers of a known size, so as one chamber is emptied, the one before it fills up, ensuring an even flow of gas at all times, while the meter registers how many times the chambers have been filled and emptied. This system is so effective that it seems to have escaped technology—the meter you use today hasn't changed much in the almost 150 years since the first one was installed in 1843.

Every month or so, someone comes around to read the dials on your meter:

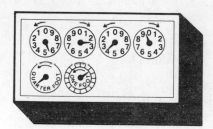

GAS METER

Reading left to right, this meter shows 5239 cubic feet of natural gas. When a needle is between numbers, always read the lower number.

Source: Pacific Gas and Electric Company. Used with permission.

The meter shows four dials: from left to right, they register cubic feet in the (1) hundred-thousands (up to 1,000,000), (2) ten-thousands (up to 100,000), (3) thousands (up to 10,000), and (4) hundreds (up to 1,000). To read the meter, read the dials from left to right, always using the smaller number when a pointer is between two numbers and ignoring the bottom dials, which are only for testing. The amount of gas you've used since the last reading is reached by subtracting that reading (the one on your last gas bill) from the current one. When your bill arrives, it will show the meter readings as a four-digit number: e.g., 5374 (prior) to 5457 (present), with a difference of 83, which means you have used 8,300 cubic feet of gas since your last bill. You may be charged for this amount, or it may first be translated into *therms.*

Therms are that latest twist in natural gas billing. *A therm is a unit of heat equal to 100,000 Btu's* (see below), and it takes 100 cubic feet of most natural gas to deliver this. But not all gas delivers the same amount of energy—some burns hotter than others—nor is all gas delivered at the same pressure, so your bill is adjusted to reflect the actual energy value you've received by using a multiplier shown on your gas bill. The multiplier is a number that translates cubic feet of gas into therms.

How many therms does it take to run appliances in your home? Home heating is usually the highest, and that one's hard to estimate, since home size, type of furnace, and insulation have so much effect on your bill. However, more and more effort is being put into making appliances energy-wise. Here are a few general estimates:

THERM ESTIMATES FOR SOME GAS APPLIANCES

Appliance	Therms
Water heater	30/month
Clothes dryer	0.15/load
Clothes washer:	
warm wash, cold rinse	0.11/load
hot wash, warm rinse	0.33/load
Dishwasher (therms for hot water)	0.16/load
Gas range:	
oven	1.33/hour
surface	1.25/hour
pilot (older models)	0.20/day
self-cleaning	0.50/cleaning
Gas log fireplace	0.20/hour
Pool heater (250,000 Btu's)	2.5/hour
Space heat (forced air, 80,000 Btu's for a 1,750 sq. ft. house)	0.80/hour furnace is on

Source: Pacific Gas and Electric Company. Used with permission.

Now, what about those Btu's?

If you've ever had to buy a furnace or a water heater, you've probably noticed that their heating capacities are given in Btu's, otherwise known as the *British thermal units.* The *Btu* does not measure temperature, but energy—by International Code agreement 1 Btu = the amount of energy needed to raise the temperature of 1 pound (lb.) of water one degree Fahrenheit. Or put another way, 1 Btu is $\frac{1}{180}$ of the energy needed to raise the temperature of 1 lb. of water from freezing (32 degrees F) to boiling 212 degrees F. The amount of energy used by appliances is usually measured in Btu's: 1 therm will provide 100,000 Btu's per hour.

☞ In countries using the metric system, heat is not measured by the Btu, but by the *small calorie*. One small calorie = the amount of energy needed to raise the temperature of one gram of water by one degree Celsius, from 14.5 degrees C to 15.5 degrees C. It takes 1,000 small calories to make the *big Calorie* (kilocalorie) used to measure the energy values of food. See Food (Energy Value), page 78. Since there are 252 small calories in 1 Btu, it takes about 4 Btu's to make one big Calorie.

GASOLINE

Do you think of gasoline much as you think of milk—squeeze the source and it all comes out the same? Even if you've wondered about that octane number on the gas pump, you've probably assumed that octane is much like cream—regular is 2 percent, premium 4 percent, and super premium is close to half-and-half—without dealing with the actual number (87 *what?*). But although a high-octane gas is considered "rich," gasolines are not the same—at this very moment each oil company may be pumping several different blends of the same octane number—meaning the same grade, such as regular unleaded—in different parts of the country. In fact, the gas you pumped today is probably quite different from the gas you got from the same pump six months ago—climate affects gasoline performance—mainly *volatility* and *anti-knock ability*—so greatly that each oil company produces variations of each grade, much as beer manufacturers vary the alcohol content of a single brand of beer to meet the different legal requirements around the country (see **Alcohol,** page 5).

Volatility. The first main performance feature of gasoline is its volatility, or tendency to vaporize. Volatility is necessary if the gasoline's to burn, but too much volatility can cause an engine to stall, while too little can leave an engine exposed to frictional heat and wear. Because volatility is greatly affected by temperature and humidity, maintaining the correct balance for different climates and seasons is tricky. The American Society for Testing and Materials (ASTM) has included five carefully defined volatility classes in its D439 Standard Specification for Automotive Gasoline. The ASTM then divides the United States into distribution areas—states or portions of states—specifying which volatility class is appropriate for each area *for each month of the year.* The volatility class applies to all the grades of gasoline.

Octane. Second is the anti-knock factor, which is where the octane number—and the grading of gas—comes into play. If you're not chummy with carburetors, "knocking" may mean no more to you than an annoying noise. Knocking—or pinging, or whatever—is much more than that, and there is good reason why oil companies push the anti-knock qualities of their gasolines. Here is what's happening: In normal combustion, a spark ignites the air/fuel mixture in the carburetor and the flame travels smoothly across the combustion chamber, with no abrupt change in pressure. (See **Motorcycles,** page 124, for information on pistons and cylinders.) If any of the fuel ignites spontaneously or from another source, like a hot spot, pressure builds up and the shock waves cause the knocking sound. A little knocking doesn't hurt much, but heavy or prolonged knocking can affect power, overheat engine parts, or even cause engine damage.

How to stop this? Although knocking may be caused by other factors, a higher-octane gasoline is often the cure. What does the octane number mean? Take this typical gasoline pump label:

MINIMUM OCTANE RATING
(R+M)/2 METHOD
87

Note that the 87 is the octane *rating,* not an actual amount of anything. The mysterious (R+M)/2 Method explains this. There are two methods for testing the anti-knock performance of a gasoline: the Research Method (R) and the Motor Method (M), resulting in two numbers: the Research Octane Number (RON), which predicts performance at low speeds, and the Motor Octane Number (MON), which predicts performance at higher speeds and temperatures and is usually higher than the RON. When these two numbers are averaged—added and divided by 2 [R+M]/2)—you get the octane rating on the label.

So what is octane? Octane, like butane or methane, is one of the

numerous kinds of hydrocarbons contained in crude oil. An isomer*
of octane, called iso-octane, is an extremely high-performance fuel
(a gasoline with normal octane molecules in it would detonate so
severely that it would ruin an engine). It would be far too expensive
to run cars on pure iso-octane, however; the octane number is an
attempt to compare the performance of a given blend of gasoline
with pure iso-octane. An octane rating of 92 tells you that that
gasoline behaves as a mixture of 92 percent octane and 8 percent
heptane (a low-performance fuel) would behave, even though that
gasoline might not have an ounce of octane in it.

How does an oil company raise or lower the octane rating? In the
early days of gasoline production—around the 1900s—refining gasoline from crude oil was fairly straightforward, without much variation in the product. But today gasoline is a blend of more than 200
hydrocarbons—molecules with different patterns of hydrogen and
carbon atoms. As developing technology allowed the industry to
"crack" large, heavy hydrocarbon molecules into smaller, lighter
ones, the volatility of gasoline could be controlled. And to improve
the anti-knock quality of gasoline, tetraethyl lead (TEL) was added—
first used commercially in 1927. In 1960 tetramethyl lead (TML)
was introduced.

When unleaded gasoline became popular, however, another
method had to be found to raise the octane rating of gasoline.
Today hydrocarbons are "cracked" and the smaller hydrocarbons
are put back together in a shape that improves the anti-knock
quality of the gasoline. This processing is more expensive than just
stirring in a lead compound.

What octane rating should you use? As a rule of thumb, use the
lowest-octane gasoline your car will run on without knocking. If,
for example, your car hums along on regular, to use premium is just
a waste of money, not to mention the extra resources required to
produce the higher grade of gasoline. Your car, however, does not
always run best on one particular octane rating. The seasons, the
place, and the altitude can change the octane rating requirements
of your car. The octane ratings in mountainous regions, for example, may be several numbers below those at sea level.

*An isomer of a certain molecule has the same number of atoms, but they are arranged
differently. Isomers of octane have 8 carbon and 18 hydrogen atoms, but the atoms have
been rearranged.

Gasoline is graded according to octane level—unleaded premium, which commonly boasts an octane rating of 91, averages about four numbers higher than unleaded regular, which is often found with an 87 rating. Leaded regular often has a slightly higher rating than unleaded regular.

Pricing. Here's a question unrelated to gasoline performance that you may have pondered: Why is gasoline priced to the last 9/10 of a cent? Isn't a 98.9-cent-per-gallon price really ridiculous? Most people who comparison shop for gasoline will buy the cheapest— the decimal columns that matter do change, and those are the ones everyone reads. This system of dividing the last penny into ten parts has been used for at least forty years because most oil companies believe that the slightly lower price sells more gasoline.

A barrel of oil. Have you ever wondered how many gallons of gasoline can be squeezed from a barrel of crude oil? From the 42 U.S. gallons of crude oil in a barrel, most processing combinations yield 8 to 25 gallons of gasoline—although 0 to 42 gallons are possible— with the balance relinquished to other petroleum products.

GOLD

The karat system is confusing, for one, because it's spelled with a "c" when referring to precious stones, but with a "k" when applied to gold. The two systems are not the same—although a one-carat diamond is fairly respectable, one-karat gold is not. When applied to gold, a karat represents a ratio of gold to alloy (other metals)— 100 percent gold is 24k, 14k gold is 14 parts pure gold and 10 parts alloy, and so on.

Why 24 parts? Why not 16 or 100? Centuries ago, gold was weighed using the troy weight system, in which a pound equaled 12 ounces (see **Weight**, page 203). A karat was used to describe ½ ounce of pure gold, or 1/24 of a troy pound.

At 24K (100 percent), gold is too soft to be made into jewelry. In the United States most gold jewelry is 14K gold—14 parts gold to 10 parts other metals, although 18K gold jewelry is also popular. By law, gold jewelry must contain at least 10 karats of gold to be labeled as karat gold. "Gold-filled" jewelry is jewelry made from a base metal with a thin layer of gold bonded to it. (It makes more sense to call this "metal-filled gold.") If jewelry is gold-filled with better than 10K gold (the minimum required for gold-filled jewelry), it is marked something like 14KGF—meaning 14 Karat Gold-Filled.

Gold can also be electroplated to metal and used in costume jewelry. The layer on electroplated jewelry must be at least seven millionths of an inch thick and at least 10K gold.

GREENWICH MEAN TIME

The U.S. military calls it "Zulu Time," pilots call it "Zebra Time," while most of the world calls it Greenwich Mean Time—the time in the Zero (also labeled "Z") Time Zone, which runs through Greenwich, England (see **Time Zones,** page 184). Greenwich Mean Time, or GMT, is a handy way to deal with time for any group that works internationally and must coordinate activities spread over many time zones—such as the military and the airlines.

GMT will always be as many hours *later* as the number assigned to your own personal U.S. Standard Time Zone: +8 for Pacific, +7 for Mountain, +6 for Central, and +5 for Eastern. So if it's 8 A.M. Pacific Standard Time, it's 4 P.M. in Greenwich. (As Greenwich Mean Time uses the 24-hour clock, it would be written as 1600 hours—see **Military Time,** page 123.) Moving from GMT to your local time—if you are in the United States, you will be 5 to 8 hours *earlier* than GMT, depending on which time zone you're in. (If you are on daylight saving time, subtract an hour from your total.)

GROSS NATIONAL PRODUCT (GNP)

Even a smidgeon of curiosity will help you through the Gross National Product (GNP). Basically, the GNP is the dollar value of all the goods and services produced in America during a specified period. The figures are reported monthly by the Department of Commerce's Bureau of Economic Analysis, with quarterly and annual summaries. We're not discussing profit here—that would be the Net National Product (see end of entry). We're talking product.

If you ran a lemonade stand for the month of July, at the end of the month you could find your gross personal product (the value of all the lemonade you sold) two ways: (1) you could calculate your *flow-of-product* by counting your take—all the money that people spent to buy your product—say $100.00; or (2) you could add your *costs* (say $40.00 for sugar, lemons, paper cups, advertising) to your *earnings* (whatever was left after your costs, which would be $60.00)—which would also equal $100.00.

The Gross National Product (GNP) does the same thing for the nation. As with the lemonade stand, there are two ways of reaching the final figure: (1) *the flow-of product method,* which adds up all the money spent to buy goods and services, or (2) *the earnings and cost method,* which adds up how much it cost to produce all the goods and services, plus the profit from their sale. Each method will add up to the same figure.

Here's what is included in the Gross National Product:

1. Consumer spending on goods and services
2. Business outlays on investments and consumer outlays for new housing
3. All government spending on goods and services (figures are broken down for federal government and state/local governments, with the federal figures divided into defense and nondefense figures)
4. Net exports (how much more—or less—we sell abroad than we buy)

It's important to note that the GNP counts only the price of *final* goods and services. Consider lasagna: The GNP is interested only in the price on the box you pull from your grocer's freezer; it does not count the price paid to the mill for the flour or the price paid to the farmer for the tomatoes, or the wholesale price paid to the factory by the grocer. The GNP counts everything only once, in the final product.

That's the good news. The bad news is that the Department of Commerce reports *two* GNPs: The *Current Dollar GNP* and the *Real Dollar GNP.* It also publishes an index of *Implicit Price Defla-*

tors. If you want to make sense of GNP figures and charts, it helps to know what these things mean.

GROSS NATIONAL PRODUCT 1979–87

(In billions of dollars; e.g., 2,508.2 = $2,508,200,000,000)

Year	Current $ GNP	Implicit Price Deflator	Real $ GNP (1982 Dollars)
1979	2,508.2	78.6	3,192.4
1980	2,732.0	85.7	3,187.1
1981	3,052.6	94.0	3,248.8
1982	3,166.0	100.0	3,166.0
1983	3,405.7	103.9	3,279.1
1984	3,772.2	107.7	3,501.4
1985	4,010.3	111.2	3,607.4
1986	4,235.0	114.1	3,713.3
1987	4,604	118.7	3,877.9

It's the job of the GNP to measure the growth of the economy—the economy's performance as compared with an earlier time. At present the year chosen with which to compare current figures is 1982. The *Current Dollar GNP* reports exactly what was paid in actual present-day dollars. But the dollar changes in value from year to year (inflation), so the Current Dollar GNP can exaggerate actual growth—when the figures rise, how do you know if we are really producing more or if the prices are simply higher? The Department of Commerce also reports how much the GNP would be in 1982 dollars. To do this, it measures how much prices increased (or decreased) since 1982—reported as Implicit Price Deflators.

The *Implicit Price Deflators* are measures of inflation—it's an index and works like any other financial index (see **Financial Indexes,** page 73), with 1982 as the base equaling 100. If prices go down, the IPD dips below 100. If prices rise, as they usually do, the IPD will be over 100. These figures can be off-putting, since the percentage symbol is never used and one doesn't normally deal with percentages over 100. But all you really need to do is subtract 100 from the figures over 100. For example, the Implicit Price Deflator for 1986 GNP was 114.1, showing that prices had risen overall 14.1 percent since 1982.

To put it all together: The 1986 Current Dollar GNP was $4,235.0 billion ($4,235,000,000,000; GNP statistics are reported in billions of dollars). In 1982 it was $3,166.0 billion. It looks like a huge growth spurt of more than a trillion dollars in just four years. But the Implicit Price Deflator was 114.1 for 1986, meaning that prices had risen 14.1 percent (14.1 above the 1982 base of 100). To correct for inflation, divide the Current Dollar GNP by Implicit Price Deflator and multiply by 100 (4,235 divided by 114.1 times 100) to get the Real Dollar (1982 Dollar) GNP for 1986 of $3,713.3.

One last word here—people sometimes confuse the Gross National Product with the Gross National Profit. The GNP is the retail price paid for everything. When you deduct what it cost to produce the goods and services, you get the *Net National Product,* which is basically gross national *profit.* It is Net National Product which interests economists in the long run, but it takes longer to collect the cost data than it does the price data, so they make do with the GNP for the short run, knowing that the NNP is usually a predictable amount less than the GNP.

HATS

Why might a man with a largish head wear a size 8 hat, while a small-headed woman would be apt to wear a size 21? Both numbers represent inches, but a man's hat size measures the hat, while a woman's measures the head. Here's how it works.

If you took the oval formed by the inside band of a man's hat, pushed it into a circle and then measured that circle's diameter, you would get the man's hat size. The perfectly round hat blocks used until about 1800 can be blamed for this awkward system—previous to that time, men's hats were made perfectly round, and men used special devices called hat screws to force their hats into a more comfortable oval before wearing them. Today men's hats are head-shaped, but the sizing system remains, in increments of ⅛ inch, the most common size being 7⅛.

Women's hats, on the other hand, are a measure of the circumference of the head that hat is designed to fit, with the measuring tape pulled just under the curve of the skull in back and across the forehead in front. Using this method, the measure of a woman's head is usually somewhere between 20 inches and 23½ inches, the most common being 22. These measures translate directly into women's hat sizes.

HEART RATE (PULSE)

Your heart has plenty of work to do, its task being to pump blood through about 100,000 miles of blood vessels. To accomplish this, the average man's heart beats 70 to 80 beats per minute, which (using the 70 figure) adds up to about 4,200 times an hour, more than 100,000 times a day. But the heatbeat of a normal human being changes from birth to old age, looking something like this:

Age	Normal pulse/minute
Newborn	as high as 110 to 160
7-year-old	about 90
Adult	60 to 80
Older people	50 to 65 not unusual

For the normal adult, the slower the resting heart rate, the stronger and more efficient the heart is considered to be, processing more oxygen with less work. The following heart rates—or

YMCA NORMS FOR RESTING HEART RATE (RHR)

Heart rates apply to adults of all ages.

Health of Heart	Percentile	RHR Male	RHR Female
Excellent	95	52	59
	85	59	63
	75	65	68
Average	50	72	73
	30	78	80
	15	84	85
Poor	5	93	92

Adapted from The Y's Way to Physical Fitness: Revised. *Used with permission of the YMCA of the USA, 101 N. Wacker Dr., Chicago, Ill. 60606.*

beats per minute—are one indication of what shape your heart is in. If you take your pulse (see below for how) while you are at rest and find your heart rate is 52, your heart is in excellent condition— among the top 5 percent (95th percentile). A male resting heart rate of 65 would put you in the 75th percentile, or in the top 25 percent of the adult hearts. And so on.

You may be concerned about putting too much stress on your heart, especially during such aerobic activity as jogging, tennis, and other sports, but it's also important to exercise at least three times a week for 20 to 30 minutes at your optimum heart rate. If your optimum heart rate is a mystery to you, you can find it—or close to it—by using this formula:

HOW TO ESTABLISH YOUR OPTIMUM EXERCISE HEART RATE

Formula	Example
1. Take the number 220	220
2. Subtract your age	-30
	190
3. Your optimum heart rate is 60% to 75% of the total in #2	
Optimum heart rate ($190 \times .70$) =	133

It's always wise to check with your doctor in matters of the heart, to find out what he or she recommends for your optimum heart rate. How do you check your pulse when you're exercising without awkward interference in your activity? You count the beats for only *ten seconds,* using a watch with a second hand, and multiply by six.

HOW TO FIND YOUR PULSE   Your pulse is caused by a stretching of the arteries that takes place after each heartbeat, most easily felt along the radial artery on your wrist. Can't find the right spot? Try this:

1. Turn your left hand palm-up in front of you.
2. Place the fingers of your right hand along the far side of the inside of your left arm, just above the thumb. (Feel the bone running along the outside of your arm? Your fingers will be just inside it.)

3. Press lightly until you feel your pulse. Use a watch with a second hand to count the beats for one minute.

Some people find it easier to find their pulse under their jaw, because the carotid artery is bigger than the radial. To find the pulse in your neck, slowly run the fingers of your left hand from your ear, past the curve of your jaw, and press them into the soft part your neck just below. Once you get the feel of your pulse, you can quickly find it again.

HIGHWAYS

There's a manual put out by the U.S. Department of Transportation's Federal Highway Administration called *Manual on Uniform Traffic Control Devices for Streets and Highways*. The introduction to this document emphasizes that the Secretary of Transportation is acting under congressional authority when decreeing that "traffic control devices on all streets and highways in each State shall be in substantial conformance with" the standards set forth. This sounds as if there were a nationally consistent system for highway markers and signs, and if you just knew what it was, you wouldn't be confused by numbers you've never been able to make sense of.

The problem is that the federal government isn't actually responsible for highway signs. That responsibility lies with what the manual admits are "a multitude of governmental jurisdictions," beginning with state governments and extending to counties and state highway districts. Although this multitude is required to pay attention to the federal manual, the reality is that "substantial conformance" ends up meaning "sometimes but not when too inconvenient." The result has been hundreds of highway labeling systems, even though there is a move to make signs more consistent throughout the country.

On the other side of it, there are some numbers that *are* part of a system, and once you see the big picture, the numbers make good sense. Here are a few of the numbered signs you might have wondered about:

DISTANCE SIGN

```
LAMAR  15
EADS   51
LIMON 133
```

Source: Federal Highway Administration, U.S. Department of Transportation.

Distance Signs. When you see this sign, are you 133 miles from the *edge* of Limon or 133 miles from the *center*! The question seems worth answering, but it turns out that the Federal Highway Administration doesn't care. Each state, or some agency within the state, sets its own rules. The result is not only inconsistency among states, but inconsistency *within* states.

For example, years ago, when California's American Auto Association was in charge of California's highway distance signs, the mileage between cities was consistently *city hall to city hall*. This system is so elegant, so sensible, that one yearns to believe it could still be true. When the State of California took over the job, however, it gave each of its twelve districts jurisdiction over highway signs, so now the mileage may be *from city limit to city limit* or *from city hall to city hall*.

So how do you know if you are umteen miles from the *edge* of town or umteen miles from the *center*! You don't.

Interstate Route Numbers. Here are some numbers that do make sense, at least much of the time. Interstate highways were originally conceived during the Eisenhower administration as a measure of national defense. Built to connect the major military installations and cities of the country (cities with a population of 50,000 or more), the name of the Interstate system was *The National System of Interstate and Defense Highways*. These highways were designed to carry large military equipment—even overpasses had to clear 16 feet to accommodate the largest missile anyone could conceive of transporting on wheels.

In the beginning Congress limited the Interstate system to 41,000 miles; later 1,500 more miles were added, bringing today's total to 42,500 miles, a figure expected to be final. Today it connects 86 percent of the major United States cities. Five of its routes

INTERSTATE SIGN   U.S. ROUTE SIGN

Source: Federal Highway Administration, U.S. Department of Transportation.

are more than 2,000 miles in length. Three run from coast to coast (I-10, I-80, and I-90), and seven run from border to border (I-5, I-15, I-35, I-55, I-65, I-75, and I-95).

The Interstate Highways route numbers are posted on blue shields with red tops and white lettering. The main *north-south* highways are always odd numbers with one or two digits. The lowest numbers begin with the West Coast's Interstate 5, increasing as you move east and ending with the East Coast's Interstate 95. Most of these north-south routes have a 5 in the route number somewhere. For example, as you move cross-country, you find routes 5, 15, 25, 35, 45, 55, 65, 75, 85, and 95, with only a few exceptions. The *east/west* highways are always even numbers with one or two digits. The lowest numbers begin in the South with Florida's Route 4, increasing as you move north and ending with the Interstate nearest the northerly Route 96. Coast-to-coast (or nearly coast-to-coast) east-west Interstates end with a zero, e.g., routes 10, 20, 40, 70, 80, and 90.

An Interstate with three digits is either a connector to or an offshoot from the main route. The first number is followed by the Interstate route number it's associated with; e.g., Interstate 580 begins and ends on Interstate 80.

U.S. ROUTES
The U.S. routes follow the same numbering system as the Interstates, only turned inside out and posted on the familiar black-on-white badge-shaped signs. The *north/south* U.S. routes are still odd-numbered, but with one to three digits. These numbers increase from east to west, starting with the most easterly, Route 1, along the East Coast to Route 101 along the West. The *east-west* U.S. routes are even numbered, but again, with one to three digits. These numbers increase from north to south, from

Route 2 along the Canadian border to Route 90 near the Rio Grande.

State and County Routes. Each state chooses its own sign to indicate the route number of state highways, and counties do the same for county roads. Generally, odd-numbered routes are north/ south and even-numbered routes are east-west, but don't depend on it. There are plenty of exceptions.

STATE ROUTE SIGNS

UTAH   NEW YORK   NORTH DAKOTA

Source: Federal Highway Administration, U.S. Department of Transportation.

Mileposts. If you ever tried to make sense of those signs that may or may not show up every mile at the side of the road, you may have given up by now. That's okay. They're not there for you. Each state installs these markers to serve mostly as reference points, used by highway patrol officers to locate accidents, by highway departments to locate road signs, and by highway maintenance crews. The green mileposts with white numbers often are readable, however; they measure the distance from either the beginning of the route or the state line. Which state line depends on the state. Although the FHA manual suggests that the numbers begin at the junctions where routes begin or at the south and west state lines, the guidelines are not always followed.

The black-on-white mileposts are even more confusing. Put in by local counties or highway districts, they often run from county line to county line, between highway district borders, or along some strange little road that begins and ends nowhere special. Trying to understand what these signs mean may be amusing if not taken too seriously. Often they measure highway maintenance districts, or distance from a county line.

Interchanges. The Federal Highway Administration favors numbering the interchanges on large highways with mileage numbers, using the interchange's distance from the state line, the beginning of the route, the beginning of the toll road, or some other reference point. Some highway interchanges are numbered in sequence; e.g., 1, 2, 3, 4, etc., making it awkward to add to a new one, say, between 1 and 2. Still, many highways don't number the interchanges at all.

HUMIDITY

When a local news report informs you that today's humidity is 95 percent, what does this really mean? Ninety-five percent of what? If the air were 95 percent water, you couldn't breathe. This "humidity" figure should always be called "relative humidity," because it does not refer directly to the percentage of moisture in the air, but to how close the air is to its saturation point, which changes with the air temperature. The relative humidity of 95 percent on a hot day probably translates into just under 4 percent moisture content. This 4 percent is as much moisture as the air can hold. If the air is cooled so the relative humidity increases to 100 percent, the air must release some of its moisture as dew, clouds, or fog.

The most accurate way to measure the humidity in the air is to extract and weigh all the water from an air sample. This process is too lengthy to be useful to most meteorologists, who depend instead on a variety of humidity-measuring instruments, called hygrometers. The oldest and most common of these, the hair hygrometer, uses one or more human hairs, because of their incredible sensitivity to humidity—human hair increases its length by about 2½ percent as the relative humidity increases from 0 to 100 percent, which explains why that nice springy hairdo goes limp on a humid day.

For an explanation of "humiditure," or the effects of humidity on the human body in high temperatures, see **Comfort Index (Weather),** page 37.

INSULATION

If you want to insulate your house and don't know your fourth R, you may want to get your feet warm, so to speak, before calling in the contractor. The language of insulation is R-value, a number that represents a material's *resistance* (hence the "R") *to heat flow.* An R-value (resistance to heat flow) equals the thickness in inches of a piece of material, divided by its thermal conductivity, or its ability to conduct heat. (Thermal conductivity is calculated using a formula involving hours, area in square feet, degrees Fahrenheit, and Btu's.) Nearly everything has an R-value—a piece of paper, a piece of cloth, your front door, a wall, and so on.

R-Values applied to insulation are regulated by the U.S. Department of Energy. The higher the R-value, or heat resistance, of a piece of material, the more effectively it will insulate. For example, the higher the R-value of the exterior walls of your house, the less heat can escape it. An exterior wall should have a minimum R-value of 11. If there is room in the walls for only three inches of insulation, you'll need to use an insulation with an R-value of at least 3.7 per inch to give you a total R-value of about 11 (3.7 times 3 inches).

Until the 1970s, the effectiveness of insulation was measured in U-values. U-values are a measure of heat *transmittance,* the opposite of R-values (which measure heat *resistance*). The less heat transmitted by a piece of insulation, the more effective the insulation will be, and the *lower* its U-value. This low-numbers-for-high-effectiveness rating system, plus the fact that the U-values for most insulation are expressed in decimals, confused people. So the government just turned the whole thing around, replacing U-values with R-values: Roughly speaking, you can divide a U-value into 1 to get the R-value. For example, a U-value of .37 is about an R-value of 2.7 (1 divided by .37).

It's important to remember R-values, especially if you're discussing your home's insulating problems with a contractor. Don't talk thickness—talk R-values. Although thickness contributes to heat resistance, the R-value is actually a combination of thickness and type of material. The Department of Energy has published several charts to help you calculate the R-values you need for your climate and amount of each type insulation that can help you reach that value.

How to Determine Your R-Value Needs

1. *Find your zone.* Most books and pamphlets use a map with wavy lines to divide the country into insulation zones. This is

confusing at best, unreliable at worst. The Department of Energy has all but eliminated this confusion by assigning insulation zones to ZIP Codes. You can identify your zone by locating the first three digits of your ZIP Code on the chart on pages 102–103. For example, if your ZIP Code is 49423, check under 494: Your insulation zone is 7.

2. Find your R-value needs. Once you know your insulation zone number, you can use this table to determine the minimum R-values your home may require. The first table applies to homes heated with gas or oil, the second to homes heated with electricity.

RECOMMENDED R-VALUES FOR EXISTING HOUSES WITH GAS OR OIL HEAT *

Insulation Zone:	1	2	3	4	5	6	7	8
Ceilings below ventilated attics	19	30	30	30	38	38	38	49
Floors over unheated crawlspaces, basements	0	0	0	19	19	19	19	19
Exterior walls**	0	11	11	11	11	11	11	11
Crawlspace walls***	11	19	19	19	19	19	19	19

RECOMMENDED R-VALUES FOR EXISTING HOUSES WITH ELECTRIC HEAT*

Insulation Zone:	1	2	3	4	5	6	7	8
Ceilings below ventilated attics	30	30	38	38	38	38	49	49
Floors over unheated crawlspaces, basements	0	0	19	19	19	19	19	19
Exterior walls**	11	11	11	11	11	11	11	11
Crawlspace walls***	11	19	19	19	19	19	19	19

*These recommendations are based on the assumption that no structural modifications are needed to accommodate the added insulation.

**For new construction, R-19 is recommended, but jamming an R-19 batt into a 3½ wall cavity will not yield R-19.

***Insulate crawlspace walls only if crawlspace is dry all year, the floor above is not insulated, and all ventilation to the crawlspace is blocked.

Source: U.S. Department of Energy.

ZIP CODES AND CORRESPONDING INSULATION ZONES*

1st 3 digits of ZIP Code	Insul. Zone	1st 3 digits of ZIP Code	Insul. Zone	1st 3 digits of ZIP Code	Insul. Zone
010-041	7	249	5	376	7
042	8	250–253	4	377–385	5
043	7	254	5	386–394	3
044	8	255–257	4	395	2
045	7	258–261	5	396–397	3
046–047	8	262	7		
048	7	263–265	5	400–407	7
049	8	266	7	408–409	5
050–053	7	267–282	5	410–418	7
054–056	8	283	5	420–422	5
056–150	7	284–285	3	423–470	7
		286	7	471	5
151	5	287–289	7	472–473	7
152	7	290–292	3	474–477	5
153	5	293	5	478–554	7
154–155	5	294–295	3		
156–158	7	296	5	556–558	8
159	5	297–303	3	559–563	7
160–167	7			564–567	8
168	5	304	2	570–571	7
169	7	305	5	572	8
170–176	5	306	3	573	7
177	7	307	5	574	8
178–185	5	308–312	3	575	7
186–195	4	313–317	2	576	8
196	5	318–319	3	577	7
197–217	4	320–326	2	580–587	8
		327–339	1	590–591	7
218	3	350–355	3	592	8
219–229	4	356–358	5	593–594	7
230–239	3	359–362	3	595–599	8
240–246	4	363–366	2		
247	5	367–369	3	600–619	7
248	4	370–374	5	620–622	5

1st 3 digits of ZIP Code	Insul. Zone	1st 3 digits of ZIP Code	Insul. Zone	1st 3 digits of ZIP Code	Insul. Zone
623	7	765	2	882	4
624	5	766–769	4	883–884	6
625–627	7	770–779	2	890–891	4
628–633	5	780	1	893–898	7
634–635	6	781–782	2		
636	5	783–785	1	900–910	1
637–639	3	786–789	2	911–912	2
640–641	5	790–791	6	913–914	1
644–646	6	792–799	4	915	2
647–658	5			916	1
660–661	6	800–815	7	917–918	2
662	5	816	8	920–921	1
664–666	6	820	7	922–925	2
667–668	5	821	8	926–928	1
669	5	822	7	930	2
670–673	5	823	8	931	3
674	6	824–828	7	932–933	4
675	5	829–831	8	934	3
676	6	832–838	6	935	4
677	7	840–847	7	936	3
678	6	850–853	2	937	4
679	5	855	4	939–954	3
681–693	7	856–857	2	955	5
		859	5	945–957	4
700–708	2	860	7	958–959	3
710–713	4	863–864	4	960	5
714	2	865	5	961	7
716–738	4	870–872	6	970–975	3
739	6			976–995	6
740–758	4	875–877	7	996–999	8
759	2	878–880	4		
760–764	4	881	6		

*If you live in Hawaii, Puerto Rico, or the Virgin Islands, your insulation zone is 1.

Source: Data from U.S. Department of Energy.

3. Look at your insulation choices. There are four main types of insulation:

Blankets or batts (rock wool or fiberglass) are fitted between exposed wood-frame studs, joists, and beams; they can be used in all exposed walls, floors, and ceilings and are good for do-it yourself jobs.

Loose fill (rock wool, fiberglass, cellulose fiber, vermiculite, or perlite) is poured between attic joists and can be used in attic floors and hard-to-reach places. This process can also be done yourself.

Blown fill (rock wool, fiberglass, cellulose fiber, or urea-formaldehyde foam) is blown into finished areas and can be used anywhere that frame is covered on both sides, such as walls.

Rigid insulation (fiberglass or extruded polystyrene board) is installed by a contractor with ½-inch gypsum board for fire safety; it is best for basement masonary walls.

How much will you need to reach the R-value goal that you established in Step 1? The government has provided you with yet another chart. This one tells you how many inches of seven types of insulation it will take to reach each of six R-values. If you want to reach R-13, for example, it will take about four inches of glass fiber batting, or 4.5 inches of rock wool batting, and so on.

INCHES OF INSULATION NEEDED TO ATTAIN SELECTED R-VALUES

R-Value Goal	Batts or Blankets		Loose Fill				
	Glass Fiber	Rock Wool	Glass Fiber	Rock Wool	Cellulose Fiber	Vermiculite	Perlite
R-11	3.5 to 4	3	5	4	3	5 to 5.5	4
R-13	4	4.5	6	4.5	3.5	6	4.5 to 5
R-19	6 to 6.5	5.25	8 to 9	6 to 7	5	9	7
R-22	6.5	6	10	7 to 8	6	10.5 to 11	8
R-26	8	8.5	12	9	7 to 7.5	12.5	9.5
R-30	9.5 to 10.5	9	13 to 14	10 to 11	8	14	11

This chart can also help you determine the value of the insulation you already have, and help you compare costs more effectively.

ISBN *NUMBERS*

You can live a meaningful life without understanding ISBN numbers—the nine numbers prefixed by ISBN on today's books. ISBN numbers are used fairly exclusively by librarians and book-sellers for ordering books. But the ISBN is interesting.

**THE BACK COVER
OF A PAPERBACK
BOOK**

Blurb describing the
irresistible contents
of this indispensable
volume!

ISBN 0-14-010654-5

The International Standard Book Number, often printed on book covers and always on the copyright page, was adopted by American book publishers in 1967 and by the International Standards Organization in 1969 to facilitate ordering by computer.

ISBN 0-14-010654-5

0: The first digit is the area of origin. 0 means that the book was published in an English-speaking country. A 4 would indicate a Spanish-speaking country, and so on.

14: The second group represents the publisher.

010654: This third group is the book's title, assigned by the publisher.

5: The last number, called the check number, alerts the computer operator to errors in the previous numbers. (See **Bar Codes,** page 7.)

LAND MEASURES

Before the Renaissance, English property was measured in *feet, rods, furlongs, acres,* and *hides,* and it still is, sort of. The rod, also known as a perch or a pole, began as an actual 10-foot pole (for the odd history of feet, see **Length, Common Short,** page 110), but by the late 1400s it was a distance of 16½ feet, and remains that to this day. The furlong, once the width of 32 plowed rows, today equals 40 rods.

It's the acre that's most familiar, since it survives as our most popular unit for estimating the size of a nice piece of land. When the word "acre" first popped into the English language, borrowed from the Latin, it merely meant a largish piece of useful farmland. By medieval times, an acre was the amount of land that a pair of oxen could plow in one day, although in some areas an acre was measured by the amount of seed required to farm it. A *hide,* no longer used, referred to the amount an ox team could plow in a year—about 120 acres.

There were obvious problems with this—a pair of aging oxen faced with heavy soil and nasty weather plowed a much smaller acre or hide than a young, feisty pair working rich soil in the sunshine. By the 1400s the acre equaled 43,560 square feet by royal decree, a measure which has survived 500 years to measure our land as well. This is how land is now measured:

Length: 16½ feet = 1 rod
           40 rods = 1 furlong (600 feet)
           8 furlongs = 1 mile
Area: 1 acre = 43,560 square feet (160 square rods)
           640 square acres = 1 square mile

A small plot of land, such as an average-size city lot, is usually measured in *square feet* instead of some part of an acre. Curiously, square yards are rarely used.

LATITUDE AND LONGITUDE

There are many persons alive today who think they know one thing for sure: that in 1492 Columbus discovered for the first time that the earth was round. But they're wrong. Columbus did help steer the world out of the medieval period of ignorance, when scientific knowledge was suppressed, but early civilizations—as far back as 5000 B.C. —*assumed* the world was round, and by 200 B.C. talk of the equator and the Northern and Southern Hemispheres was common. A librarian named Eratosthenes even measured the equator 2,200 years ago by using the sun, a well, a vertical column, and some camels. He came up with a figure of 28,700 miles, truly in the ball park (the correct measure is 24,902 miles).

Erastosthenes is known today in mapping circles as the father of geodesy—the science of earth measurement. He was the first to chart the earth with parallel east-west and north-south lines, although he did not place them at regular intervals. Hipparchus of Nicaea soon did, making a grid of 360 north-south lines that went from pole to pole and 180 lines circling the earth, each parallel to the equator. (See **Circles,** page 30.)

Hipparchus's *north-south lines* ran from the North Pole to the South Pole and were called *meridians,* a word that meant "noon," for when it is noon on one part of the line, it is noon at any other point as well. Meridians were used to measure *longitude,* or how far east or west a particular place might be, and were 70 miles apart at the equator, making the equator 25,200 miles long, only 298 miles off the mark.

The *east-west* lines were called *parallels,* and unlike meridians, were all parallel to each other and measured *latitude,* or how far north or south a particular place might be. There were 180 lines circling the earth, one for each degree of latitude.

The system has survived for the most part—the meridians are now about 69 miles apart at the equator and the parallels are 69 miles apart, but the lines in both directions are still separated by one degree. (See illustrations on page 108.)

THE WANDERING PRIME MERIDIAN Although meridians and parallels appeared on maps for hundreds of years, there was a

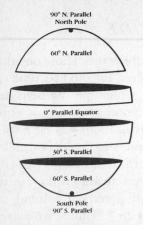

PARALLELS slice the earth like an onion.

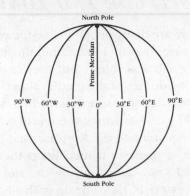

MERIDIANS section the earth like an orange.

perplexing problem: When numbering the lines of longitude, where did one start? This was no problem for latitude—the equator was the obvious choice for 0° latitude. But where did one put the Prime Meridian, or 0° longitude?

Until an amazingly short time ago—just over 100 years—the location of the Prime Meridian was a matter of patriotism—each country simply ran it through its own capital, numbering its maps accordingly. The Prime Meridian has run through Paris, Madrid, Cracow, Copenhagen, Rome, Augsburg, Beijing, St. Petersburg, Washington, and the Royal Observatory at Greenwich, England. In 1884 the first International Meridian Conference finally put the Prime to rest at the Royal Observatory in Greenwich, England, where it has remained to this day. At last, time zones could be assigned to the world, impossible without an agreed-upon Prime Meridian. (See **Time Zones,** page 184.)

All the lines could now be numbered the same way on all maps, and coordinates for all places in the world would be standard.

READING COORDINATES To understand how to use coordinates, look at a globe. Longitudes east of the Prime Meridian (running through England) are numbered 1° E to 179° E (180° is the halfway point—near the International Date Line); those west are lettered 1° W to 179° W. Likewise, latitudes north of the equator are numbered 1° N to 90° N (the North Pole); those south of the equator are numbered 1° S to 90° S (the South Pole). Latitudes are easy to find, as they are marked on the metal meridian. Longitudes are more difficult, usually marked along the equator. (Note that not all the meridians and parallels are put in—that many lines would obscure the printing.)

THE WESTERN WORLD BY LATITUDE AND LONGITUDE

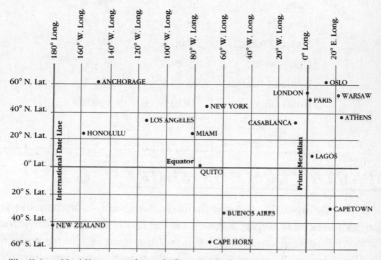

The Prime Meridian runs through Greenwich, England.

Each degree of latitude or longitude is divided into 60 minutes and each minute can be divided into 60 seconds. Usually minutes are symbolized by an apostrophe (60′) and seconds with ditto marks (60″). Sometimes, though, they're written like a decimal; however, 11.46 N, a latitude reading, is *not* a decimal, as the number after the "decimal point" indicates minutes, so it's never higher than 59. The same goes for seconds.

The coordinates are two sets of numbers—one for longitude and one for latitude. Use the longitude reading first to find the correct meridian, and line it up with metal meridian, the guide bar found on all globes. Now read the latitude on the metal bar and bingo! You're there!

Official coordinates for international locations are given in some atlases. Look for some of these places that you've probably heard of but might not be sure where they really are:

Place	Longitude	Latitude
Chad	17.48 N	19.00 E
Fiji	18.40 S	175.00 E
Tasmania	38.20 S	146.30 E
Caspian Sea	40.00 N	52.00 E
Transylvania	46.30 N	22.35 E
Paraguay	24.00 S	57.00 W
Grenada	12.02 N	61.15 W
Midway Island	28.00 N	179.00 W
Gobi Desert	43.29 N	103.15 E
Falkland Islands	50.45 S	61.00 W
Costa Rica	10.30 N	84.30 W

LENGTH (COMMON SHORT)

Most people assume that the American system of measuring the length of everyday things is based on the foot. But exactly how long *is* a foot? Twelve inches, yes, but how long is an inch? ¹/₃₆ of a yard? But how long is the yard? These are not stupid questions. Perfectly intelligent people have been asking them for hundreds of years, and none of the many answers has been without problems.

Two of the earliest measurements were the width of a man's thumb for the inch and the length of his foot for the foot. The tools couldn't have been handier and the system was user-friendly, but people with different-sized feet couldn't do business. So in Egypt official measuring sticks the length of the king's foot were distributed. However, each new ruler changed all the measures, a frustrating custom that spread northward through Europe. Finally, in the fourteenth century, King Edward I set a standard by producing a yard based on his ample girth. Yardsticks were made and circulated. But standards continued to change (it was once changed to the distance from Queen Elizabeth's nose to the end of her longest finger), and scientists began demanding more uniformity and precision.

At last France proposed the metric system in 1971, based on the distance of a line running from the equator, through Paris, to the North Pole. A meter, calculated to be one ten-millionth of this distance, was set into platinum and kept at Sèvres, France, at the International Bureau of Standards. By the beginning of this century, most of the world used the metric system.

But even the meter was in trouble. Truly exacting scientific measurements required something more accurate than a length of metal that could change with time and temperature. In 1960 the meter was changed to equal 1,650,763.73 wavelengths in a vacuum of the orange-red line of the spectrum of the gas krypton-86. Although the krypton-86 system was accurate to one ten-millionth of an inch, it was not useful for lengths over 8 inches, so it was changed again. Today's meter, official since 1983, is measured by laser and is equal to the length of the path traveled by light in a vacuum during a time interval of 1.299792468 of an atomic second. This meter can be reproduced in science laboratories.

Congress based our American yardstick on the meter, so when the meter changes, our yardstick changes with it. For general use, however, the yard remains equal to 0.9144 of a meter. (In 1959, a change made the new yard shorter by two parts in a million, a difference so insignificant for general use that the old equivalent is still used.) (See also **The Metric System,** page 116.)

There is a way to spell relief from these tediously exact measures: i-n-c-h. The inch is by no means sacred, and although it's been customary for centuries to use binary (based on 2) fractions such as ½, ¼, ⅛, etc., there is nothing official about it and nothing special about sixteenths. However, when you're using a pocket calculator, decimal inches are far easier to work with than the awkward fractions presented by more traditional inch divisions. It's becoming more and more common to divide the inch into tenths, called a *decimal inch.* Industries usually employ the decimal inch (the ones not yet gone metric), and machinists routinely measure their work in thousandths of an inch, or *mils (1 mil = ¹⁄₁,₀₀₀ of an inch).* The thickness of plastic bags, for example, is measured in mils—a 2-mil garbage bag is about ¹⁄₅₁₂ of an inch by the fractional system. Microinches are used to measure the roughness or smoothness of a surface.

LIGHT BEER

Did you ever wonder just how many calories was in that can of "light" beer you've been drinking? You may have looked for this information on the can or bottle and found nothing at all.

Federal regulations do not answer the question, but do provide some general guidelines: In order for a company to call a beer "light," it must have one-third fewer calories than the same company's regular beer. Regular beers average about 150 calories; light beers average about 100 calories. That does help some, except that the calories in beer vary so much that one company's regular may have only 5 or 10 more calories than another's light.

How do the beer companies cut the calories in light beer? Mostly by cutting the alcohol content. Alcohol, which is highly caloric, accounts for about two-thirds of the calories in regular beer; cutting the alcohol content of a beer is going to cut the calories. As a result, light beers usually contain a lower percentage of alcohol than their regular counterparts. If you're a light drinker, the 50 or so calories—the amount contained in a few nuts or small pretzels—you may save by switching to a light beer may not be worth the sacrifice in taste. Light beers often taste blander than regular beers.

LIGHT BULBS

Until recently you probably never read the packages that incandescent light bulbs came in except for the large-print wattage, which is how light bulbs are sized. You put a 40-watt bulb in the closet, a 60-watt bulb in the hall, and a 100-watt bulb next to your reading chair, and that was that. Then, despite the fact that you had never complained about light bulbs, the companies that made them began to complicate one of the few remaining simple chores in American life by offering you "improvements."

Now you can buy "long-life" bulbs, "extended service" bulbs, and krypton "energy-saver" bulbs. To make the choice even more difficult, companies you never heard of are offering bargain-priced bulbs, leaving you unsure of what you're getting. If you really want to know, you have to start reading the package.

Every light bulb package gives at least three pieces of information: wattage, lumens, and life expectancy. The *wattage* is not brightness, but a measure of the electricity you'll pay for when you use the bulb (see **Electricity,** page 60). Average *lumens* is the brightness. One standard candle produces one lumen; one standard 100-watt light bulb usually produces about 1,600 lumens. The *life expectancy* of the bulb is given in number of hours.

For comparison's sake, a casual perusal of several hardware light bulb departments produced these lumens and life expectancy figures:

APPROXIMATE LUMENS AND HOURS:
Standard Light Bulbs

Watts	Lumens	Life Expectancy
15	125	2,500 hours
25	215	2,500 hours
40	480–510	1,000 hours
60	880	1,000 hours
75	855–1,210	750 hours
100	1,600–1,750	750 hours

APPROXIMATE LUMENS AND HOURS:
Long-Life Light Bulbs

Watts	Lumens	Life Expectancy
40	about 460	1,500 or more
60	610–820	1,300–3,500 hours
75	820–1,125	1,000–3,500
100	1,350 to 1,630	1,000–3,000 hours

When you compare bulbs, you'll find there are trade-offs—you may sacrifice brightness (lumens) and a bargain price to get a longer life. Bargain-priced bulbs may be so low on lumens that a 100-watt bulb may not be much brighter than a standard 75-watt bulb. Lumen for lumen, standard bulbs are probably still a better buy than long-life bulbs, no matter what the wattage of the bulb. However, the convenience of longer-lived bulbs may also be attractive—you don't need to change them as often. Bargain-priced bulbs, on the other hand, really can be a bargain if the life expectancy is not reduced and the slightly dimmer (sometimes more yellow) light is tolerable to you.

It's interesting to note how much the use of a light bulb really costs you. A 100-watt bulb will burn for 10 hours for the price of 1 kilowatt-hour of electricity (1 kilowatt-hour runs 1 watt for 1,000 hours). If it burns out after 750 hours, that bulb cost you $7.50 in power (if your electric company is charging you 10 cents per kilowatt hour) plus the cost of the bulb. Light bulbs may use only about a penny's worth of power an hour, but the cost adds up fast. (Fluorescent bulbs are an alternative for large areas, giving two to three times the lumens per watt of power than incandescent bulbs.)

LUMBER

Because there are so many ways to grade lumber, depending on type, size, and/or use, few people can be expected to understand them. Redwood alone is graded by the California Redwood Association into 6 basic grades with more than 50 offshoots. Hardwood is graded differently from softwood, "shop" grades differ from construction grades, and so on.

However, there are two things that might help you when deciding what type of lumber to buy. The first is understanding the difference between *linear feet* and *board feet*. There is great variation in what kind of lumber is sold in board feet and what's sold in linear feet, but all you really need to know is how to figure each of them out:

Linear Feet. If a board is priced at $1.00 per linear foot, you will be charged $1.00 for every foot of that board that you buy, no matter what size the board is. This is as easy as it sounds. Surfaced lumber (partially finished, not rough-sawed) is often sold by the linear foot.

Board Feet (BF). Rough lumber is often sold by the board foot. A board foot is a piece of wood measuring 1 inch thick, 12 inches long, and a foot long, i.e., a square foot 1 inch thick. Most lumber sold by the board foot is not a convenient 1 inch thick and 12 inches wide, but you still need to figure out the board feet if you want to know how much the lumber will cost you. Here's the formula:

$$\frac{\text{thickness in inches} \times \text{width in inches} \times \text{length in feet}}{12}$$

This really works. Try it: you have a board 1 inch thick, 6 inches wide, and 8 feet long priced at $.50/BF. What will it cost you? Set up in the formula, it looks like this:

$$\frac{1\,(\text{thickness in inches}) \times 6\,(\text{width in inches}) \times 8\,(\text{length in feet})}{12}$$

$$= \frac{48}{12} = 4 \text{ board feet} = \$.50/\text{BF} = \$2.00 \text{ cost for the board}$$

The second thing you should understand is that *all lumber is measured in the rough,* so if you buy a 2 × 4 in rough lumber, it will measure 2 inches by 4 inches. If you buy it "surfaced," it will lose some of the original wood in the finishing process, making it measure more like 1½ × 3½ inches. The rule of thumb used by many who frequent lumberyards is this: When you buy surfaced lumber, you normally lose about ¼ inch (for 1-inch boards) to ½ inch (for 2-inch boards) in thickness and about ½ inch in width.

Now, about those grades. "Clears" are boards with no knots and are usually quite expensive. "Select" boards are of a lesser quality lumber than clears, but the defects are minor, with defects or knots under two inches.

Construction lumber is often graded as follows:

2-Inch-Thick Lumber Only	Description
No. 1 (Construction)	Many knots, all under 2 inches
No. 2 (Standard)	Many knots, up to 3½ inches
No. 3 (Utility)	Allows open knots, splits, pitch
No. 4 (Economy)	Lowest grade

Most other lumber is subject to a complex grading system with each type of lumber—ponderosa pine, redwood, cherry, etc.—graded differently. The "Standard Specifications for Grades of California Redwood Lumber" is over 100 pages long. It's no wonder the novice woodworker gets confused.

THE METRIC SYSTEM

Little wreaks more havoc on the average American nervous system than the threat of metrics, officially called the System of International Units (SI). You thought that the United States would never switch to the Metric System, didn't you—that in the 1970s when metrics roared, grassroots America rose up and chased it off? Our holdout is hurting our wallets—more and more countries are refusing to buy our nonmetrically sized and packaged goods—and wallets usually have the last word. Whether you are aware of it or not, you have been edging, inching, even centimetering toward metrics.

But it's not as bad as you think. You're already somewhat familiar with metrics. For starters, consider these things you deal with fairly automatically:

Item	Metric Measure
Electricity	Watts, volts, amp, ohms
Radio frequencies	Hertz
Food	Calories
Jewels	Carats
Soda pop and wine	Liters
Tar, nicotine, sodium cholesterol, many vitamins and medicines	Milligrams
Many tools and fasteners, blood pressure, cameras, film, binoculars	Millimeters
Engine capacity	Cubic centimeters

Actually, your relationship with metrics goes even deeper than that—you've been using metrics secondhand all your life. When you measure with yards, pounds, and gallons, you are actually using converted meters, kilograms, and liters—by Act of Congress, our weights and measures have been based on the following metrics for almost a hundred years.

THREE BASIC METRIC MEASURES

Length. Although the metric system sounds modern, it's not new. The French made the first meter in 1791, calculating its length as a romantic one ten-millionth of a line running from the equator through Paris, to the North Pole. A platinum meter was established with great care in a suburb of Paris at the International Bureau of Weights and Measures. Subsequently, more sophisticated measurements of the earth proved the first meter incorrect, so in 1889 a new bar, which was the length of the first but not based on any earthly measure, was substituted. But a bar can change minutely and is difficult to replicate with complete accuracy, so in 1960 scientists defined the meter as equaling the transition between the levels $2 p_{10}$ and $5 d_5$ of the krypton-86 atom. Now the meter could be measured exactly in any properly equipped laboratory.

Today's meter, officially accepted in 1983, is equal to the length of the path traveled by light in a vacuum during the time interval of $1/299\ 794\ 458$ of a second. (See **Length [Common Short]** page 110.) Our yard has been officially defined as equal to $3600/3937$ of a meter, whatever its definition, since 1893. For general use, 1 yard equals 0.9144 meter.

Mass (called weight in U.S. measures). The kilogram was first defined as the mass (weight) of one cubic decimeter of water at the temperature of maximum density. Scientists needing a more precise measure voted in 1875 at a General Conference on Weights and Measures to replace this with a cylinder of platinum-iridium alloy that was the same size but not based on anything, and was kept at the International Bureau of Weights and Measures. Our copy of the official kilogram, kept at the National Bureau of Standards, is the basis for our avoirdupois pound (the one you weigh yourself in), which officially equals 0.45359237 kilogram, or for general use, 0.453 kilogram.

Volume. The liter is not really a basic measure, since it is derived from the kilogram. It originally was the volume occupied by a kilogram of pure water. In 1964 the liter was redefined as a cubic decimeter of pure water, a measure that differed from the first by only 28 parts in a million. This new standard is important only in high precision measurements. The U.S. quart is equal to 1.101 liters.

THE METRIFICATION OF AMERICA

The little-known truth is that the metric system has influenced U.S. measures since the year after the end of the Civil War, when its use was made legal in this country (1866). In 1893 the meter and kilogram officially became the basis of our weights and measures. Not much happened for a very long time after that until President Gerald Ford signed the current metric bill in 1975 and set up the U.S. Metric Board that was supposed to facilitate a switch from U.S. measures to metrics. That bill made the change voluntary, however, so little really happened—opposition was so heated that politicians feared for their seats. In 1982 President Reagan abolished the board.

But metrics marches on. We are the last English-speaking holdout—Britain, Australia, New Zealand, and even Canada have switched to the metric system. The only nonmetric country in the world besides ourselves is Burma. Our dearly beloved system of weights and measures is becoming a national embarrassment and an economic liability. In 1988 President Reagan signed a bill that in part mandates all government agencies to go metric by 1992. We are talking about metrics in your lifetime.

You may feel out of control when you lose your gut feeling for the size and number of things. However, if metrics were the only system around, you'd quickly develop new gut feelings. American resistance to metrics is so passionate that you now have to cope with *both* systems. What could be more awkward? You have to speak metric but you think in United States. What you once figured in your head now requires a calculator. This mess is caused by our foot-dragging, not because the metric system is difficult; in fact, nothing could be simpler.

PREFIXES

Ten is the magic metric number—each power of ten is assigned a prefix, which can be attached to nearly any metric measure. This elegant simplicity, combined with the ease of metric math, has made the International System of Units the chosen system of the world. (For very high and very low metric measures, see **Prefixes, Astronomical,** and **Prefixes for the Minuscule,** page 141.)

Prefix	Meaning	Examples
kilo-	= 1000	kilogram, kilometer, kiloliter
hecto-	= 100	hectogram, hectometer, hectoliter
deka-	= 10	dekagram, dekameter, dekaliter
(no prefix)	= 1	gram, meter, liter
deci-	= 0.1 ($\frac{1}{10}$)	decigram, decimeter, deciliter
centi-	= 0.01 ($\frac{1}{100}$)	centigram, centimeter, centiliter
milli-	= 0.001 ($\frac{1}{1000}$)	milligram, millimeter, milliliter

The following common objects can help give you a feel for metric measures:

Item	Approximate Weight
A paper clip	1 gram
A dollar bill	1 gram
A dime	2 grams
A penny	3 grams
A nickel	5 grams
A quarter	6 grams
A 2-pound can of coffee	1 kilogram

Item	Approximate Length
A "fat yard" (a yard + a hand)	1 meter
The width of a paper-clip wire	1 millimeter
The width of a dime	1 millimeter
A sugar cube	1 cubic centimeter
The width of the littlest fingernail	1 centimeter

Item	Approximate Volume
A "fat quart" (1 quart + ¼ cup)	1 liter

Until metric measures are the only measures used in America, however, you'll have to depend on conversion charts, like a foreign language dictionary. The awkward translations are probably to blame for the widespread misconception that the metric system is difficult.

U.S./METRIC AND METRIC/U.S. This information is included to give you a feeling for sizes; although you can use it to convert from one measure to another, the Conversion Factors that follow make quicker work of that chore.

1 acre	=	0.405 hectare
1 bushel	=	35.238 liters
1 centimeter (cm.)	=	0.39 inch
1 foot	=	0.31 meter
1 inch	=	2.54 centimeters
1 gallon	=	3.785 liters
1 grain	=	0.0648 gram
1 gram (g)	=	0.035 ounce
1 hectare (ha.)	=	2.47 acres
1 kilogram	=	2.2046 pounds
1 kilometer (km.)	=	0.62 mile
1 liter (l, or L)	=	0.908 dry quart 1.057 liquid quarts
1 meter (m)	=	39.37 inches
1 mile	=	1.609 kilometers or 1,609 meters
1 milligram (mg.)	=	0.02 grain
1 milliliter (ml.)	=	0.06 cubic inch
1 millimeter (mm.)	=	0.04 inch
1 ounce	=	28.349 grams
1 (fluid) ounce	=	29.573 milliliters
1 (dry) pint	=	0.550 liter
1 (fluid) pint	=	0.473 liter
1 pound	=	0.453 kilogram
1 (dry) quart	=	1.101 liters

1 (fluid) quart	=	0.946 liter
1 (long) ton	=	1.02 metric tons
1 (metric) ton	=	2,204.62 pounds
1 (short) ton (2,000 lbs.)	=	0.907 metric ton
1 yard	=	0.914 meter

CONVERSION FACTORS

This fearsome chart is probably what killed metrics in the 1970s, but now you've got what it takes—a calculator.

When You Know This Unit	Multiply by This Number	To Get
acres	0.40	hectares
bushels	35.2	liters
centimeters	0.4	inches
	0.328	feet
cups	0.24	liters
cubic feet	0.028	cubic meters
cubic yards	0.765	cubic yards
cubic meters	35.3	cubic feet
	1.31	cubic yards
feet	30	centimeters
	0.3	meters
gallons	3.8	liters
grains	64.8	milligrams
grams	0.0353	ounces
hectares	2.47	acres
inches	2.54	centimeters
	25.4	millimeters
kilograms	2.2	pounds
kilometers	0.6	miles
kilometers per hour	0.621	miles per hour
liters	4.2	cups
	2.1	pints
	1.06	quarts

When You Know This Unit	Multiply by This Number	To Get
liters	0.26	gallons
meters	3.3	feet
	1.1	yards
	0.000621	miles
	0.00054	nautical miles
miles, statute	1.6	kilometers
	1600	meters
miles, nautical	1862	meters
miles per hour	1.609	kilometers/hour
milliliters	0.2	teaspoons
	0.067	tablespoons
	0.034	fluid ounces
millimeters	0.04	inches
ounces (fluid)	30	milliliters
ounces (dry)	28.3	grams
pints	0.47	liters
pounds	0.45	kilograms
quarts	0.95	liters
square centimeters	0.155	square inches
square feet	0.0929	square meters
square inches	6.45	square centimeters
square kilometers	0.4	square miles
square meters	1.2	square yards
square yards	0.836	square meters
tablespoons	15	milliliters
teaspoons	5	milliliters
tons, short (2,000 lbs.)	0.9	metric tons
tons, metric	1.1	short tons
yards	0.9	meters

How to get rid of these annoying, terrifying, time-consuming charts? The answer is as simple as metrics itself—switch!

MICROWAVE OVENS

When you read an ad for an .8 cubic foot microwave oven, it's hard to visualize how big that really is— anything behind a decimal point sounds minuscule. Microwave ovens are sized by the cubic foot capacity of the interior, and a cubic foot is bigger than you may think.

WHAT A MICROWAVE OVEN WILL HOLD

.4 cubic foot	a loaded salad plate
.6 cubic foot	a piled-high dinner plate
.8 cubic foot	a 5-pound roast
1.6 cubic feet	an 18-pound turkey

The exterior measurements vary from brand to brand:

.4 cubic foot	approx. 18″ wide × 9″ high × 15″ deep
.6 cubic foot	approx. 18″ wide × 11″ high × 15″ deep
1.4 to 1.6 cubic feet	approx. 25″ wide × 14″ high × 18″ deep

The power for a microwave oven ranges from 500 watts (low) to 720 watts (high). Low-watt ovens cook more slowly than higher ones; e.g., a 500-watt oven takes about 8 minutes to bake a potato, while a 720-watt oven takes about 4 minutes.

MILITARY TIME

One of the cultural surprises awaiting the new military recruit is the 24-hour time system—does 1900 hours mean anything to you? Still, the 24-hour time system makes a lot more sense than ours. Times showing up twice a day can be confusing—to indicate which side of noon a time falls on, it must be followed by the clumsy A.M. or P.M., which is frequently left off. That this does not cause more confusion than it does is due only to our sense of what is appropriate. If your boss invites you to his house to meet the family, saying, "Drop by at seven," you're unlikely to show up for breakfast.

124 *MOTORCYCLES*

The military takes time seriously, however. Not only are its operations spread over many time zones, the appropriate time for what one is called on to do may not always be obvious. So military time divides the day not into two sets of twelve hours but into one set of twenty-four hours, counting from 1:00 A.M. our time (0100, or "oh-one-hundred") to midnight (2400, or "twenty-four-hundred"). This not only gets rid of the awkward Latin qualifiers but the punctuation as well:

1:00 A.M. becomes 0100

2:10 A.M. becomes 0210

Noon becomes 1200

1:00 P.M. becomes 1300

3:00 P.M. becomes 1500

6:00 P.M. becomes 1800

9:54 P.M. becomes 2154

Midnight becomes 2400

This very sensible system might have caught on with civilians if clocks numbered 1 to 12 hadn't been around so long. In fact, official national and international time is always communicated in 24-hour time, with many other groups—pilots, for one—using it as well. It's not hard to translate 24-hour time into familiar time, though. The A.M. times are obvious. For P.M. subtract 1200 from numbers larger than 1200; e.g., 1200 from 1900 is 700, or 7 P.M. (See also **Greenwich Mean Time,** page 89.)

MOTORCYCLES

Because so many gas-powered engines traditionally have been measured by horsepower, many people don't know that the number on a motorcycle is *not* a measure of horsepower—most 1000cc bikes produce less than 100 horsepower. (See **Engines [Horsepower],** page 63.)

The number on the side of a motorcycle means cc's, or cubic centimeters. Cubic centimeters of what? Unfortunately, although the answer does concern engine size, it's not so simple as "the volume taken up by the engine." Cubic centimeters, as applied to a motorcycle engine—or any other internal combustion engine—is a measure of the combustion space in the cylinders.

The first internal combustion engine, built in 1876 by German engineer Nicholas Otto, contained the essential parts of today's gasoline engines:

1. a cylinder
2. a piston, pumping up and down inside the cylinder
3. a crankshaft connected to the piston
4. valves to admit air and fuel into the cylinder at the beginning of the cycle and to let the exhausted gases escape at the end
5. a device to explode the fuel-air mixture; e.g., a spark plug

This engine, like today's engines, worked in four strokes of the piston in the cylinder:

1. The *intake stroke:* The piston moves down, sucking the fuel-air mixture into the cylinder.
2. The *compression stroke:* The piston goes up to compress the fuel-air mixture into a small space.
3. The *power stroke:* The fuel-air mixture explodes (combusts), pushing the piston back down.
4. The *exhaust stroke:* The piston returns to the top, pushing the exhausted gases out of the cylinder.

Although Otto's engine couldn't be used for transportation—it had to be hooked up to gas lines—by 1882 another German engineer, Gottlieb Daimler, invented an ingenious device called a *carburetor.* The carburetor turned liquid gasoline into a spray so fine that the drops evaporated immediately, mixing with the air sucked up by the piston's *intake stroke,* and exploding easily on the *compression stroke.*

Today's engines, though greatly refined, work essentially the same way (although some now have a 2-stroke cycle instead of a 4-stroke). The size of an engine is measured by theoretically removing the top of a cylinder, pushing the piston all the way down, and then filling the cylinder with liquid. The cubic centimeters of liquid displaced (spilled out) when the piston is returned to its high position is the measure of the combustion area of that cylinder. If a motorcycle has four cylinders, each displacing 200 cc's, it has an 800cc engine (200 cc's times 4 cylinders).

The formula for finding the displacement in cc's in a cylinder is as simple as finding the volume of a tin can: You simply find the area of the circle at the top (see **Circles,** page 30) and multiply it by its length:

$$\text{pi} \times \frac{(\text{bore})^2}{2} \times \text{the stroke} = \text{displacement}$$

It may help to define the terms:

pi	=	3.1416
bore	=	diameter of the cylinder in centimeters
stroke	=	the distance the piston moves in centimeters
displacement	=	volume of a cylinder in cubic centimeters (cc's)

Automobile engines are sized essentially the same way.

NAILS

The sizing of common nails is one of those deliciously archaic systems that have survived miraculously intact, with today's sizes running from 2d to 60d. The "d" is the British symbol for pence, which is why a 2d nail is called a "2-penny" nail. In 1400, a 2-penny nail was a size that sold 100 nails for 2 pence, or 2d. Likewise, 100 3-penny nails sold for 3d, 100 16-penny nails for 16d, and so on. Today most commonly used nails are sold by the pound (weight), not by the count, but they are still called "penny" nails. The most commonly used household nails are common nails (large, flat head), box nails (like common nails but thinner), casing nails (cone-shaped head), and finishing nails (small head and thinner than casing nails), all of which come in penny sizes. Other nails are often sized simply by length in inches.

Although standards sometimes vary, the following chart provides fairly good guidelines for the length for the 16 most popular nail sizes. Note that the lengths increase ¼ inch for each size under 16d and ½ inch for each size over 16d.

Size	Length	Size	Length
2d	1″	10d	3″
3d	1¼″	12d	3¼″
4d	1½″	16d	3½
5d	1¾″	20d	4″
6d	2″	30d	4½″
7d	2¼″	40d	5″
8d	2½″	50d	5½″
9d	2¾″	60d	6″

The gauge, or the diameter of the nail, doesn't usually concern users of the penny nails, as it simply increases with the length of the nail. Technically, however, nails use a 15-1 gauge system, the diameter of the nail increasing as the gauge numbers go *down* (a 10-gauge nail is thicker than a 12-gauge nail). This odd system is loosely related to wire gauges, as most nails are cut from wire—the thickness of wire also increases as the gauge numbers decrease, running from the largest at 00000 (½-inch diameter) to a small 40 (.007 inch). Confusion reigns, of course, because the nail gauge system works the opposite way from the screw system, in which the screw width increases as the gauge numbers go *up* (see **Screws and Bolts,** page 156).

How do you choose a nail size? A rule of thumb used for many purposes is that a nail driven through the thickest part of two pieces of wood should be long enough to go through both, less ³⁄₁₆ of an inch. When edge-nailing—driving a nail lengthwise along a board—the nail should be three times longer than the thickness of the first piece. Although there are plenty of exceptions to the rule, it's a place to start if you're an amateur staring glassily at the nail bins.

An interesting bit of nail history is that until 1800, nails were made by hand. The first ones were probably made from bone and wood, and when the Iron Age came along, from metal. The wooden nails—called *treenails*—stuck around, however, used even in Colonial America, when metal nails were so precious that it was not uncommon for people to burn their houses when they moved, in order to have enough nails to build the next one! Nails were hoarded, with old ones being straightened to use again. George III insisted that nails be imported from England, where they were still being made by hand—making wrought-iron nails had become a household industry, to fill the incredible demand, and by 1740, some 60,000 people were making nails in the English town of Birmingham alone! The American habit of self-reliance dealt such a blow to the English economy that in 1750 the English passed the Iron Act forbidding Americans to build their own ironworks; we could not even make our own nails. Shortly after, the Revolutionary War broke out. By 1800, our factories were making cut nails. The "French nail"—the nail made from wire—soon replaced cut nails and persists to this day.

Cut nails were cut from sheet metal and were rectangular or square—you can still buy square nails, but they tend to be a specialty item. Today's more common round shape became widely used when machines began making the nails from wire.

OIL (ENGINE)

Like most people, you probably have assumed that the "w" in 10w/40 car engine oil stands for "weight." Wrong. Engine oil with two viscosity ratings, such as 10w/40, is simply oil that has been tested at both cold and hot temperatures. The "w" stands for "winter." Basically a car oil needs to be thin enough to flow in winter so you can start the engine, but thick enough when the engine gets hot (and in areas with high summer temperatures an engine can get very hot) so that the oil can perform its main function of keeping the metal surfaces apart. Its ability to do this is called its "film strength."

Oils with two ratings ("multigrade" oils) contain added plastics with long chain molecules called polymers. These flexible molecules allow a thick oil with good film strength in summer to remain thin enough to allow the engine to start in winter. (Oils with just one rating have been tested only at high temperatures, and would have to be changed to suit the season.) The rating numbers are arbitrary numbers assigned by the Society of Automotive Engineers ("SAE" on the oil can). The 10 isn't 10 *of* anything; it simply stands for oil of a certain thinness when tested at 0 degrees.

Oils tested at 0 degrees are rated 5, 10, 15, or 20. Oils tested at 210 degrees are rated at 20, 30, 40, or 50. What do those numbers mean? To assign a number, the SAE times in seconds a 60cc sample of oil as it runs through an opening similar in size to a standard carburetor jet.

> *"W" ratings* (oil tested at 0 degrees F):
> —10w oil runs half as fast as 5w oil.
> —15w oil runs half as fast as 10w oil.
> —20w oil runs half as fast as 15w oil.

> *Regular ratings* (number of seconds a 60cc oil sample takes to run through a test aperture):
> —20 oil takes 45 to 58 seconds.
> —30 oil takes 58 to 70 seconds.
> —40 oil takes 70 to 85 seconds.
> —50 oil takes 85 to 110 seconds.

How to choose the oil for your car? Check the owner's manual; most makers recommend an oil. As a rule of thumb, the lower your winter temperatures, the lower the "w" rating needs to be, and the higher the summer temperatures, the higher the regular rating.

PAPER

What does "20-pound" mean when applied to paper? And a ream is 500 sheets of 8½-by-11-inch stock, isn't it? There are many types of commercial paper: When a ream (using a stack of 500 sheets of paper measuring 2 feet by 3 feet) weighs 20 pounds at a temperature of 70 degrees and a humidity of 50 percent, it's called 20-pound paper. This is how some commercial paper is sized. The heavier the weight, of course, the thicker the paper usually is, but the pound designation always applies to that oversized stack, *not* the 500 8½-by-11-inch sheets you buy at the stationer's.

A ream is not always 500—sometimes it's 480 sheets. And the size of the sheet is not always 2 feet by 3 feet—sometimes it's 17 inches by 22 inches. Usually you can get 2,000 letterhead sheets out of a ream, but not always.

It's usually at the printer's that you have to choose the weight of your paper. Despite the confusion of size and number of sheets in a ream, here are a few general guidelines:

SOME COMMON COMMERCIAL PAPER WEIGHTS

9 lbs.	onionskin
16 lbs.	mimeographing paper
20 lbs.	standard typing paper
24 lbs.	standard letterhead paper
60 lbs.	good for printing on both sides
65 lbs.	business cards, postcards
100 lbs.	magazine paper (coated stock)
120 lbs.	poster board

In addition to commercial paper, art paper is also weighed, but is sized by a different standard. The size of the sheets in the stack to be weighed (again 480 to 500 sheets) depends on what the paper is for. The standard size for watercolor paper is 22 inches by 30 inches, but other art papers may not be the same:

SOME COMMON ART PAPER WEIGHTS

8 to 48 lbs.	tracing papers
13 to 20 lbs.	bond papers
28 to 32 lbs.	ledger papers
65 to 80 lbs.	cover papers (commercial weight)
90 to 400 lbs.	watercolor papers

PAPER CLIPS

Paper clips are a snap, but deserve some explanation, because not only are the size numbers backward—the bigger the number, the smaller the clip—but one size seems to be missing (#2), while another has no number at all. "Gem clips" is the standard title for the familiar U-shaped wire, which comes in two widths: .040 and .036 inch.

GEM CLIP SIZES

#3	small
#1	standard
	jumbo

PENCILS

You'd think that the pencil would just fade away, what with pattering keyboards and ubiquitous ballpoint pens. Pencil popularity, however, seems here to stay—more than 2½ billion pencils are produced in America every year. The U.S. government uses 45 million of them a year, and the New York Stock Exchange more than a million. Perhaps it's because the pencil's an old-fashioned hard worker—one standard pencil can leave a 35-mile trail, or about 45,000 words. At least forty materials from twenty-eight countries go into one.

What does the No. 2 on the pencil signify? It's a measure of hardness. About 1800 a Frenchman named Nicolas Jacques Conté mixed clay and water with powdered graphite, rolled thin the result, and baked it in a 1900-degree kiln. He encased the dried vermicelli-like "leads" in a wood sandwich, inventing the first modern pencil.

Today numbers and letters indicate the hardness or softness of a pencil lead, or to put it another way, how much clay is mixed with the graphite to produce the lead—the more clay in the mixture, the harder the lead. Sizing pencils would be a simple matter, except that there are several kinds of pencils—art pencils, writing pencils, and mechanical pencils, and each kind uses a different system.

ART PENCILS Art pencils have a wide range of hardnesses to accommodate the many artistic purposes pencils serve, from drawing to drafting. The scale goes from the soft end—8B—to the hard end—10H. Sizes 8B through F are often used for artwork, while 3H through 10H are preferred for drafting. It works like this:

"B" means soft; the higher the number that precedes "B," the softer the pencil, and the wider and darker the line produced.

"H" means hard; the higher the number that precedes "H," the harder the pencil, and the thinner and lighter the line produced. "F" means fine, but not as fine as the H numbers. Observe:

8B-7B-6B-5B-4B-3B-2B-B   **HB-F-H**   **2H-3H-4H-5H-6H-7H-8H-9H-10H**

Softest <<<<<<< Softer   Middling   Harder >>>>>>>>>> Hardest

WRITING PENCILS Writing pencils don't require such a wide range as art pencils, and tend to fill that gap in the middle. Writing pencils use numbers 1 through 4 to indicate hardness, which relate directly to the art pencil system:

Writing pencil		Art pencil
1	=	3B
2	=	B
2.5	=	F
3	=	2H
4	=	3H

MECHANICAL PENCILS You don't have to sharpen a mechanical pencil, and often you can choose the kind of lead you'd like. However, there isn't much choice. Measured in millimeters (diameters), the most available sizes are 0.3 mm., 0.5 mm., 0.7 mm., and 0.9 mm. Hardnesses follow the art pencil system.

pH

From time to time you may run into a pH value, a number which indicates how acid or alkaline something is. On a scale of 0–14, pH values are applied to soils, "acid rain," waste, paper, fruits, soaps, etc. The scale, though not difficult, is not particularly obvious. For example, you might assume that the lower the pH, the weaker the acid; instead, the opposite is true—*the lower the pH, the more acidic the solution.*

Technically, "pH" describes the concentration of free hydrogen ions in a solution. (The letters "pH" mean "potential of hydrogen.")

You don't have to know about hydrogen ions to use the pH scale, but for the curious, here it is: The pH number is not a count of ions, but a "gram-equivalent," or weight in grams per liter of solution. As it happens, the concentration of hydrogen ions in a liter of solution can range from a 1 gram-equivalent to a .00000000000001 or 10^{-14}, gram-equivalent. In 1909 a scientist named Soren Sorensen used this natural spread to come up with a 0–14 scale. On the scale 0 was the most acid, or the strongest concentration of hydrogen ions (1 gram-equivalent), 7 (pure water) was neutral (10^{-7} gram-equivalent), and 14 had the lowest concentration of hydrogen ions, making it the most alkaline at 10^{-14} gram-equivalent.

How much stronger is beer at a pH of 4.0 than black coffee at 5.0? The pH scale is logarithmic—each whole number represents a concentration ten times stronger than the next higher number. So the beer would contain ten times the number of hydrogen ions as the coffee. To demonstrate further: A lime with a pH of 2.0 is ten times more acidic than an apple with a pH value of 3.0, which is in turn ten times more acidic than a tomato with a pH of 4.0 (making the lime 100 times more acidic than the tomato), and so on.

THE pH SCALE

pH Value	Hydrogen Ion Gram Equivalent	Relative Acidity or Alkalinity	
0	1	10,000,000	MOST ACID
1	10^{-1}	1,000,000	
2	10^{-2}	100,000	
3	10^{-3}	10,000	
4	10^{-4}	1,000	
5	10^{-5}	100	
6	10^{-6}	10	ACID
7	10^{-7}	1	NEUTRAL
8	10^{-8}	10	ALKALINE
9	10^{-9}	100	
10	10^{-10}	1,000	
11	10^{-11}	10,000	
12	10^{-12}	100,000	
13	10^{-13}	1,000,000	
14	10^{-14}	10,000,000	MOST ALKALINE

These approximate pH ranges for things in daily life may hold some surprises. Were you aware, for example, that apples are as acidic as grapefruit? Most of our food and our body products, for that matter, with the exception of our blood, is on the acidic side. The following list is ordered by approximate average pH value:

SOME AVERAGE pH VALUES

Range or Average pH	Example
0.1	Hydrochloric acid (normal)
0.3	Sulfuric acid (normal)
1.0 to 3.0	Stomach acid
1.6	Limes (fruit)
2.0 to 5.6	Acid rain
2.0 to 4.0	Soft drinks
2.3	Lemons
2.4 to 3.4	Vinegar
2.8 to 3.8	Wines
2.9 to 3.3	Apples
3.0 to 3.3	Grapefruit
4.0 to 5.0	Beers
4.5 to 5.5	Human hair
4.5 to 5.5	Shampoos "best for hair"
4.5 to 4.7	Bananas
5.0	Black coffee
5.0 to 7.0	Human urine
5.6 to 6.0	Normal rain
6.3 to 6.6	Most drinking water
7.0	Pure water
7.3 to 7.5	Human blood
7.6 to 8.0	Eggs
8.0	Seawater
9.0	Toilet soap
8.4	Sodium bicarbonate (0.1 normal)
10.6 to 11.6	Ammonia
14.0	Sodium hydroxide (normal)
14.0	Drain cleaner

Gardeners are particularly interested in the pH values of their soil, since plants can be very sensitive to too much acid or alkaline (see **Soil, Garden,** page 166).

PINS

If you sew a lot, you probably vacuum up plenty of pins to prevent one from lodging itself in a family foot. Before the early 1800s, however, the heads of pins were attached by hand, an expensive process, so losing one in the carpet would have been unthinkable. In fact, in the 1100s pins were so valuable that the English Parliament passed a law allowing them to be sold only on January first and second of any year.

Even in Colonial America, pins were scarce. Workers cut the pins from brass wire and crimped on the heads separately from a finer piece of wire, using a foot-powered drop hammer. Not surprisingly, the heads were prone to fall off, and elaborate measures were taken to keep pins from getting lost. By 1860, however, pins were made by machine at the incredible rate of 1600 per hour, not much by today's standards, perhaps, but enough to change everyone's attitude toward them for good.

Today most pins are made of stainless steel with brass or nickel coating. Pins appear to be the simplest of sewing accessories until you are presented with the bewildering array of them offered in some fabric stores. Pins are even used in banking to hold securities and tissue-thin receipts. Although even the people who sell you pins don't seem to know too much about them, it's worth getting the right ones for the job.

PIN TYPES Widths of pins vary not so much by length, but according to their intended use, from extra-fine silk pins to sturdy quilting pins. Some pins have plastic or glass heads to make handling easier. Although these pins work well for pinning patterns, they often get hit by sewing machine needles. If you pin fabric for sewing with ball-headed pins, pull them as you sew.

SOME COMMON PIN TYPES

Type of Pin	Size	Width	Use
Satin	16–18	Fine to medium	All around pin for fabrics that handle easily
Silk	24	Extra-fine	For silks and other slippery fabrics
Pleating	16	Fine	Holds set pleats for sewing without breaking machine needles
Quilting	28	Large	Strong for holding thick or many-layered fabrics, pile

PIN SIZES Pins are sized by the sixteenth of the inch. A pin measuring one inch, for example, is a size 16, or $^{16}/_{16}$ of an inch. A 1¼-inch pin is a size 20, or $^{20}/_{16}$. On the small end, a size 6 pin is $^{6}/_{16}$ of an inch, or $^{3}/_{8}$".

COMMON PIN SIZES

Size	Length	Size	Length
6	$^{3}/_{8}$"	20	1¼"
8	½"	24	1½"
14	$^{7}/_{8}$"	28	1¾"
16	1"	32	2"
17	$1^{1}/_{16}$"		

PLYWOOD

Although most people think of plywood as a fairly modern invention, it was actually invented by the ancient Egyptians who, around 2800 B.C., fastened plies of wood together with wooden pegs. There was a difference, however: Wood was rare in Egypt, but today wood, and especially plywood, is common and found in many types and grades.

Plywood sizes and grading are fairly straightforward compared to many other types of lumber. The thicknesses are measured in fractions of an inch, and the sizes of the panels are measured in feet. Unlike surfaced lumber, for example, a ¼-inch-thick 4-by-6-foot panel actually measures ¼ inch thick, 4 feet wide, and 6 feet long—

and plywood is priced by the panel, so no elaborate figuring is required. You want to buy the least expensive plywood to do the job right, however, so you need to know the letters.

The letters you see most often applied to plywood—A, B, C, or D—refer to its "grade," or quality, but other letters are sometimes found:

A	knots are plugged and sanded
B	larger knots, but plugged and sanded
C	open knots
D	larger open knots
G	good side (for finishing)
I	for interior use*
S	side
SH	sheathing (rough on both sides)
WB	wallboard (sound face, utility back)
U	underlayment (for floors), sometimes UL
X	for exterior use

*The I is often omitted; if it isn't marked for exterior use (X), it's intended for interior use.

These letters are combined with each other or with numbers. For example:

G1S	one good side, one utility side
G2S	both sides good
ACI	one side A, one side C, for interior use
BCX	one side B, one side C, for exterior use

POINTS, MORTGAGE

It goes without saying that when you shop for a mortgage, you look for low interest rates, but when the subject of points is raised you may very well draw a blank.

Actually, it's a fairly simple system—the points on a loan are *prepaid interest.* Each point will cost you 1 percent of your mortgage amount, so a mortgage for $80,000 with 2 points will cost you an extra $1,600 ($800 per point). You may pay this amount when you sign for the loan, or the amount may be added to the total amount of the mortgage. Points are not always charged to the buyer, however; they can be negotiated so the seller pays some or all of the points. In the case of VA or FHA loans, the buyer can be charged only 1 point—the seller pays any remaining points.

How you choose between a loan at, say, 12½ percent interest costing 2 points or a loan for 12¾ percent interest costing 1 point depends on how long you intend to keep the property. Each point effectively raises your interest ⅛ of 1 percent on a 30-year mortgage, so the first loan will raise the effective interest rate to 12¾ percent, a lower rate than the 12⅞ percent of the second loan. If you plan to keep the property for a long time, the first loan will cost you less in the long run, because your total interest rate is lower. However, if you intend to keep the property only a short time, you don't want to pay any more interest up front than you have to, so the 1-point loan might be a better choice.

Points can fluctuate so fast that a few days can cost you a bundle. Why? Because money is a commodity and is as volatile as the stock market. When money is in short supply and the economy is weak, points go up so that banks will have more available money up front. To keep borrowers from being scared off by high interest rates, lenders may "discount" (charge points on) a loan—the interest will be lower, but the points ensure the lender of a good "yield," or profit. A loan with no points, called a *par loan,* is likely to carry a fairly frightening interest rate.

POSTAL RATES

Because the Post Office classes mail from first class through fourth, you probably have assumed that the system makes sense. First-class mail is faster and costs more than third-class (books, circulars, etc.), but at this writing, third class is slower and costs less than fourth (parcel post). Strangely enough, you can send a package under 2 pounds first class (delivery time 1 to 5 days) for only 10 cents more than fourth class (delivery time 1 to 3 weeks), if it's going coast to coast. If the destination is local, however, fourth class might save you more than a dollar. What happened to second class? It's mainly for newspapers and magazines "with second-class mail privileges."

It's worth asking before you pay, or, better yet, get the free flyer that the Postal Service calls *Postal Rates, Fees and Information* that charts the rates in detail. (Postal rates change so often that the rates quoted here may already be out of date!) Although the following rates may have changed by the time this book emerges from production, they are included to give you a sense of the differences in rates for different services.

U.S. POSTAL SERVICE RATES

1ST CLASS MAIL: charged by the ounce
- 1st ounce .... 25 cents
- Each additional ounce up to 12 ounces .... 20 cents

1ST CLASS ZONE RATED PRIORITY MAIL: charged by the pound (2-day delivery, usually)
- Up to 2 pounds, any zone $2.40
- Over this, check Postal Rates flyer or Post Office

POSTCARDS (Single) .... 15 cents

PARCEL POST is charged by the pound and by distance traveled to one of eight zones. For mail over 1 pound, rates start at $1.43.
Consult Postal Rates flyer or your Post Office for other rates.

EXPRESS MAIL (Next-day service): charged by the pound
- First ½ pound $8.75
- ½ to 70 pounds .... rates vary by weight and zone

INTERNATIONAL AIRMAIL up to ½ ounce .... 45 cents
- Canada (up to 1 ounce) .... 30 cents
- Mexico (up to 1 ounce) .... 25 cents

Source: U.S. Postal Service, April 1988.

Although postal rates do seem to be getting very high, until 1851 (the Second Continental Congress appointed the first Postmaster General, Benjamin Franklin, in 1775), the cost of sending a single-sheet letter 40 miles was either 6 cents or 8 cents. To send it 400 miles cost 25 cents. These prices were doubled or tripled or more with each additional sheet of paper. These could easily be counted, since envelopes were not used and the sheets were simply folded with the address on the outside sheet. Mail had to be picked up from the local post office—it cost extra to have it delivered. No packages were delivered until 1912 when Parcel Post became law.

Those delivery prices were expensive for the early days, but with few roads and slow transportation, it was no easy task to transport a letter to its proper destination. As the Post Office expanded, how-ever, the rates came down. Only now have postal rates again reached 25 cents per letter—although we can send more than one sheet!

HISTORY OF FIRST-CLASS MAIL RATES

Year	Cents per First Ounce	Year	Cents per First Ounce
1885	2	1971	8
1917	3	1974	10
1919	2	1975	13
1932	3	1978	18
1958	4	1981	20
1963	5	1985	22
1968	6	1988	25

PRECIOUS STONES

Precious stones are measured in *carats,* a measure of weight that equals about $1/142$ of an ounce. This strange number is the weight of a locust tree seed ("carat" comes from an Arabic word meaning "bean" or "seed") which the Arabs once used for weighing gems and gold. Today the international carat weighs 200 milligrams, or $1/5$ of a gram, and is a fairly respectable size; a 1-carat diamond would be about the size of a dried pea. Frequently, diamonds displayed in ordinary jewelry are not as large as that.

PREFIXES, *ASTRONOMICAL*

Before 1958, when the International Committee on Weights and Measures gave us even larger measures, a mega-something was the biggest something anyone could think of. And why not? *Mega-* is a prefix meaning *million.* Millionaires made megabucks, megahertz measured radio waves, and a megaphone impressively amplified sound.

Unfortunately, mega- can't begin to be used to measure our galaxy, much less the universe, so now we have *giga-* and *tera-* as well:

kilo-	thousand (1,000)
mega-	million (1,000,000)
giga-	billion (1,000,000,000)
tera-	trillion (1,000,000,000,000)

All these prefixes, although they sound terribly scientific, are rather charmingly based on Greek synonyms for "really large": *kilo-* comes from the Greek word for "thousand," *mega-* from "great," *giga-* from "giant," and *tera-* from "monster." Although the prefixes are commonly used with metric measures, making teragrams and gigameters and such, it's quite acceptable to attach them to nonmetric measures, making kilowatt-hours and megatons. There is an exception: Metric prefixes have a slightly different meaning in the world of computers (see **Computers,** page 43). It shouldn't be long before these handily descriptive prefixes become common usage for "more than the mind can comprehend," with billionaires making "gigabucks," rock stars using "gigaphones," and "teradollars" describing the National Debt.

PREFIXES FOR THE MINUSCULE

However far scientists have reached into the universe, they have reached even farther into the minuscule, and particle physicists require even more prefixes than astronomers. Hardly anything is more difficult to comprehend than the unseeable structure of the visible world. Still, you're probably not bothered by things "microscopic," since you can see them through a microscope. *Micro-,* meaning *millionth,* has come to mean "incredibly small, about as small as you can get," with spies hiding messages on microdots and libraries reducing vast card catalogs to books of microfilm.

But *micro-* in today's known invisible world is an unwieldy prefix. Here are some others:

milli-	thousandth	(0.001)
micro-	millionth	(0.000001)
nano-	billionth	(0.000000001)
pico-	trillionth	(0.000000000001)
femto-	quadrillionth	(0.000000000000001)
atto-	quintillionth	(0.000000000000000001)

(To keep from confusing these with prefixes meaning "huge," remember that prefixes for the smallest numbers end with *-o,* while, except for *kilo-,* those for the largest numbers end with *-a.*)

The dividing prefixes are usually applied to metric measures, like meters, grams, even seconds. An exception is the inch, which can be divided into microinches. (*See* **Length** [**Common Short**], page 110.) Although seconds are multiplied by 60 to make hours, when seconds are divided, they go metric, splitting into milliseconds, microseconds, even picoseconds. What could possibly be so small? To get an idea, make a dot by pressing the point of a pen or pencil on a piece of paper. This dot measures about 1 millimeter. It's small, but at least you can see it! Shrink the dot to 1 micrometer and you could still see it with a good optical microscope. But at .1 micrometer (100 nanometers) it would disappear, being smaller than the wavelength of visible light.

It's hard to believe that things as small as the following can be so accurately measured:

• An amoeba, which can be seen through an ordinary optical microscope, weighs about 5 micrograms.
• Wavelengths of visible light are measured in nanometers.
• A liver cell weighs about 2 nanograms.

- Many atoms measure from 100 to 600 picograms.
- Subatomic particles are often measured in femtometers.
- Viruses must be measured in attometers.
- An electron weighs 0.00000000009 attogram.
- A neutrino, the smallest known mass, weighs
  0.00000000000005 attogram!

So how small is small? If you have a virus that measures 1 atto-meter, you could sling 1 quadrillion of those viruses (that's 1,000,000,000,000,000) across your pencilled dot! Now, if you were to enlarge each of those quadrillion viruses to the size of your millimeter dot and line them up end to end, they would stretch 1 billion kilometers (or 620,000,000 miles) into space, almost reaching Saturn!

PRIME RATE

In times of economic excitement, news of prime interest rate changes—say from 9 percent to 8.75 percent—has jumped from the business page to the front page of many newspapers. One-quarter of 1 percent may not sound like much, but the repercussions roll quickly—sometimes within hours—across the country. You probably know that when the prime rate goes up or down, other interest rates—including home mortgages—usually go up or down with it, involving, on a national scale, enormous amounts of money.

The prime rate is the lowest rate of interest commercial banks offer—it's the rate charged for short-term loans to customers with impeccable credit ratings. Unless you're worth millions, you're un-likely to get anywhere close to these rates. Mere home mortgages or small business loans will always be higher.

Why is there just one prime rate when there are so many banks? Although all banks may not always offer the same prime rate, the prime rate generally tends to become standard across the banking industry. When one major bank—a big one like Security Pacific, Bank of America, Wells Fargo, New York's Chase Manhattan, or Citibank—lowers its rate, the others often follow so quickly that you wonder how such enormous organizations can move that fast. Because the cost of nearly everyone's money can move up and down with it, the action is tracked by small investors as well as the very large.

Since 1971 the prime rate has changed frequently—sometimes weekly—from a low of 5 percent in the early 1970s to 21½ percent in late 1980. The prime rate did not always move around so much. Before 1971 the prime rate was administratively fixed. It was the First National City Bank of New York that led a move to a flexible— or floating—rate that reflects what's happening in the money market and at the Federal Reserve, where commercial banks, large corporations, and other wealthy borrowers get *their* short-term loans, usually at lower interest than the prime rate. So although mortgages and smaller loans are not directly tied to the prime rate, they tend to follow its ups and downs.

PRODUCE

Supermarket produce is sometimes advertised by size, such as size 60 avocados or size 18 artichokes. The secret is this: *The bigger the size number of the produce, the smaller the actual size will be.* Many kinds of produce are sized by how many items fit into the shipping box. This means that size 60 avocados are shipped to the market 60 to an avocado box; thus, size 60 avocados are quite small. Size 18 artichokes, on the other hand, come only 18 to a box, so they will be quite large.

Unfortunately, there are so many variables that one number does not really stand for small, medium, or large. Although the box size is the same for each kind of produce, it may differ from other kinds—apple boxes are always the same size and orange boxes are always the same size, but an apple box is not quite the same size as an orange box. (Many boxes are sized to carry 50 pounds of produce.) Then, apples need packing material to keep them from bruising, but oranges usually don't. Finally, there's the obvious difference in fruit sizes—a medium-sized avocado is about the size of a small artichoke.

Here are a few examples of produce sizes to give you an idea of how the sizes work (sizes can vary widely, of course):

SOME ESTIMATED PRODUCE SIZES

Type of Produce	Small	Medium	Large
Apples and oranges	120	80–88	56–64
Artichokes	48	36	18
Avocados	60	48	24
Potatoes		100	70–80
Tomatoes (often bought by the pound)		48	24

PROPERTY (LEGAL DESCRIPTION)

Surveyors' legal descriptions of the exact whereabouts and boundaries of property, or "parcels," often found on deeds, appraisals, contracts of sale, and in county records, can be intimidating. It may help, however, to know some early surveyor's measures, as many older property descriptions still remain:

7.29 inches	=	1 link		
25 links	=	1 rod	=	16½ feet
4 rods	=	1 chain	=	66 feet
10 chains	=	1 furlong	=	200 yards
8 furlongs	=	1 mile	=	32 rods	= 5,280 feet

These odd units are based on Gunther's chain, a seventeenth-century English surveyor's tool that was 66 feet long and was composed of 100 links joined by small rings. The 66 feet fit in neatly with the rod and furlong measures used for centuries in England. More primitive materials were also used for surveying—rope, leather—even hair, twisted, stretched, and dried, was popular in early American days. The steel surveyor's tape, first made in Massachusetts from ladies' hoopskirt wire about 100 years ago, succeeded the chain—and the hair—for most purposes. Today sophisticated compasses and sighting instruments find and measure the boundaries of property, while most of the old measures have given way to *decimal feet*—the foot is divided into tenths instead of inches (twelfths).

Unfortunately, there are three ways to survey property, all of them legal and all of them still in use. A property description may use the Metes and Bounds method, the Rectangular system, or the Lot and Block system, but it's not unusual to find two or more in the same description.

Metes and Bounds. This system is the oldest, most commonly used, and tedious to read. The property is located by its proximity to local landmarks and neighboring property (bounds) to find a place of beginning—often using another survey method to do so. From there, the surveyor walks the perimeter of the property, measuring angles in degrees and minutes and the distances between them. The following property description finds the point of

METES AND BOUNDS SURVEY

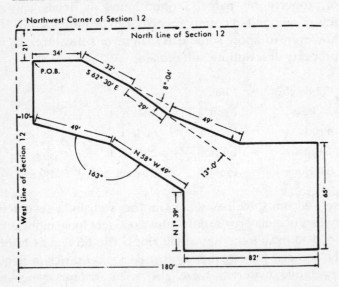

Used with permission from The Appraisal of Real Estate, *9th edition.* © *1987 by the American Institute of Real Estate Appraisers, Chicago.*

beginning by using the Rectangular survey system (explanation follows), then proceeds with Metes and Bounds.

Commencing at the Northwest corner of Section 12 thence South along the Section line 21 feet; thence East 10 feet for a place of beginning; thence continuing East 34 feet; thence South 62 degrees, 30 minutes East 32 feet; thence Southeasterly along a line forming an angle of 8 degrees, 04 minutes to the right with a prolongation of the last described course 29 feet; thence South 13 degrees, 0 minutes to the left with a prolongation of the last described line a distance of 49 feet; thence East to a line parallel with the West line of said Section and 180 feet distant therefrom; thence South on the last described line a distance of 65 feet; thence due West a distance of 82 feet; thence North 1 degree West 39 feet; thence North 58 degrees West a distance of 49 feet; thence Northwesterly along a line forming an angle of 163 degrees as measured from right to left with the last described line a distance of 49 feet; thence North to the place of beginning.

The Rectangular System. The Rectangular system was adopted by the Second Continental Congress, when it needed to sell off parts of its huge land acquisitions quickly—staking out each perimeter with a chain and compass would have painfully held up the much-needed revenue. To avoid this, the U.S. General Land Office established reference points—true east-west lines, individually named, called *base lines,* and north-south lines, also named, called *principal meridians.*

Land to be surveyed was, and still is, divided by north-south lines, six miles apart, called *range lines,* and east-west lines, also six miles apart, called *township lines.* Range lines are numbered east and west from the principal meridian, and township lines are numbered north and south from the base line. When lines intersect, they make 36 square-mile *townships.* Each township is divided into 36 *sections,* each a mile square, or about 640 acres. (Due to the curvature of the earth, adjustment lines also run every 24 miles, making some sections inexact.)

A legal description of a 20-acre parcel located in the highlighted section on page 148 begins with the site and ends where the principal meridian and base lines intersect: The west half of the northeast quarter of the southeast quarter of Section 10, Township 4 North, Range 3 East, Mount Diablo Base and Meridian.

RECTANGULAR SURVEY

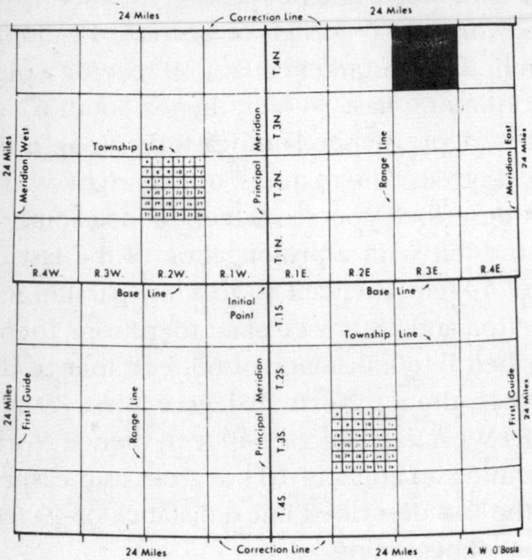

Source: John S. Hoag, Fundamentals of Land Measurement, Chicago Title Insurance Company, 1976. Used with permission.

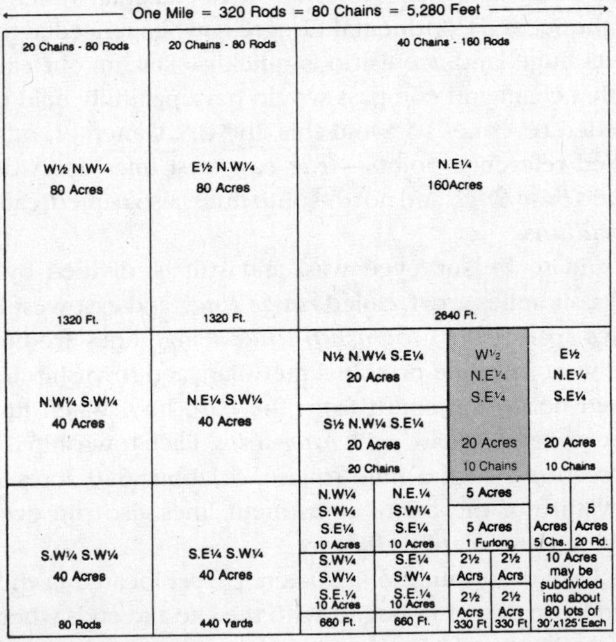

One Mile = 320 Rods = 80 Chains = 5,280 Feet

20 Chains - 80 Rods	20 Chains - 80 Rods	40 Chains - 160 Rods		
W½ N.W¼ 80 Acres	E½ N.W¼ 80 Acres	N.E¼ 160 Acres		
1320 Ft.	1320 Ft.	2640 Ft.		
N.W¼ S.W¼ 40 Acres	N.E¼ S.W¼ 40 Acres	N½ N.W¼ S.E¼ 20 Acres / S½ N.W¼ S.E¼ 20 Acres / 20 Chains	W½ N.E¼ S.E¼ 20 Acres 10 Chains	E½ N.E¼ S.E¼ 20 Acres 10 Chains
S.W¼ S.W¼ 40 Acres	S.E¼ S.W¼ 40 Acres	N.W¼ S.W¼ S.E¼ 10 Acres / S.W¼ S.W¼ 10 Acres / S.E.¼ S.E.¼ 10 Acres	N.E.¼ S.W¼ S.E¼ 10 Acres / S.E¼ S.W¼ 10 Acres / S.E.¼ S.E.¼ 10 Acres	5 Acres / 5 Acres 1 Furlong / 2½ Acrs 2½ Acrs / 2½ Acrs 2½ Acrs 330 Ft
80 Rods	440 Yards	660 Ft.	660 Ft.	5 Acres 5 Chs. / 5 Acres 20 Rd. / 10 Acres may be subdivided into about 80 lots of 30'x125' Each

Used with permission from The Appraisal of Real Estate, 9th edition. © 1987 by the American Institute of Real Estate Appraisers, Chicago.

The Lot and Block System. This method is often used by land developers to subdivide a parcel. Usually a surveyor lays out the streets; then lot lines are agreed upon by the owners, each lot being assigned a number. Three systems can be at work here: the parcel being developed might be located by the Rectangular system, the lots identified by the Lot and Block system, and the exact property lines might be further established by Metes and Bounds.

LOT AND BLOCK SURVEY

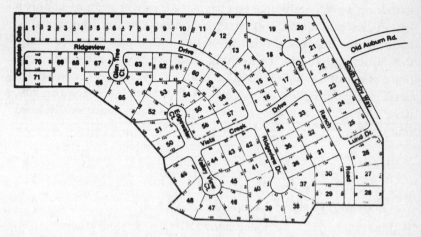

Used with permission from The Appraisal of Real Estate, *9th edition.* © *1987 by the American Institute of Real Estate Appraisers, Chicago.*

Is this confusing? You bet. Still, though you may not be able to write a legal description, it's nice to know what's generally going on when you sign those intimidating documents for your dream house.

RADIO WAVES

Did you ever wonder why the AM numbers on your radio dial are bigger than the FM numbers? Or what the difference was between regular (VHF) television channels and UHF channels? Or why you sometimes hear a CB radio in the middle of your favorite rerun? In fact, what do these things, plus electricity, microwaves, infrared waves, light waves, X rays, and gamma rays, have in common? All are electromagnetic waves, all of which travel at the same speed— the speed of light—and each of which vibrates at a constant rate.

What makes one electromagnetic wave different from another is how fast it's vibrating, or the *frequency* (number) of the waves, called *cycles,* that go by per second. Frequency is measured in *hertz:* 1 hertz = 1 cycle per second. Very low frequency waves with long wavelengths, like electricity (AC power) vibrate at only a few cycles per second: 60 hertz is common in the United States. Radio waves begin at about 15,000 hertz. Compared to electricity, that sounds high, but it's nothing compared to X rays, which vibrate at about 1,000,000,000,000,000,000 cycles per second (10^{18} hertz) or gamma rays at more than 10^{24} hertz (see **Exponents,** page 64). Hertz are also referred to in larger, more easily used units:

HERTZ

1 cycle per second	= 1 hertz (Hz)	
1,000 hertz	= 1 kilohertz (kHz)	
1,000 kilohertz	= 1 megahertz (MHz)	= 1,000,000 hertz
1,000 megahertz	= 1 gigahertz (GHz)	= 1,000,000,000 hertz

How does this translate to your radio dial? The AM side, usually numbered from 550 to 1600 (some dials remove the last zero, leaving it 55 to 160), stand for *kilohertz,* although today's AM band extends from 525 to 1,700 kilohertz, or 525,000 to 1,700,000 cycles per second. The FM side of your dial is usually numbered from 88 to 108, which stands for *megahertz.* FM numbers are lower than AM numbers, but the frequencies are much higher— 88,000,000 to 108,000,000 cycles per second. The FM stations are sandwiched between television stations, which are assigned frequencies according to channel: VHF channels 2 through 6 broadcast at 54 to 88 megahertz, below the FM frequencies, while channels 7 to 13, broadcasting at 174 to 216 megahertz, and the UHF channels (14 to 83), broadcasting at 470 to 890 megahertz, are above the FM channels. CB radio uses two bands, one of which is in

460 to 470 right under the UHF band, which accounts for its occasional television interference.

Whatever it's broadcasting, each station is assigned its frequency by the FCC—the Federal Communications Commission—which has been regulating American broadcasting since 1934, to keep stations from interfering with each other. Each station operates strictly within its assigned channel, whose size depends on the type of broadcast. AM channels require only a 10 kilohertz band, while FM channels are closer to 200 kilohertz and television channels require 6,000 kilohertz each.

So how big is a radio wave? The length of a wave (cycle) is measured from crest to crest, or from the tip of one wave to the tip of the next. Very low frequency waves (lower than 30 kilohertz) can measure over 10,000 yards—more than six miles—from crest to crest. Medium frequency waves—AM broadcasting waves fall in here—are about 1,000 to 100 yards each. VHF waves—used for FM and television broadcasting—measure 10 yards to 1 yard. UHF waves are from about a yard to half an inch. Extremely high frequency waves, such as X rays, are so small that they are measured in Angstroms (one ten-billionth—0.0000000001—of a meter): light rays are approximately 3,900 to 7,700 Angstroms wide, while an X ray might measure 1 Angstrom, and gamma rays can be smaller than 0.000001 Angstrom.

It's the size of the electromagnetic wave, related to its frequency, which is really what makes a light wave (which you can see) different from an electrical wave, or a radio wave, or an X ray. The range is phenomenal—frequencies run from 1 to more than 1,000, 000,000,000,000,000,000,000 hertz, with wavelengths measuring from several miles to far smaller than 0.0000000000000001 meter.

RAIN

There is hardly a more misunderstood weather forecast than the prediction of rain. If you read that there's a 50 percent chance of rain for today, it sounds as if the weather people were guessing that there's half a chance that the area covered by the forecast (yours) will get rained on, as if they were betting 50-50 odds. Although many local forecasters still hang on to this idea, the National Weather Service insists that the forecast percentage is *not a probability,* but the percentage of your area that *will be wet* by the end of the day. One could argue that odds are still 50-50 that you won't need your umbrella.

Rain reports—or the measure of rainfall in inches—represent the amount of rain that would have remained on the ground if it did not run off or soak in. To calculate this, rain gauges—straight-sided containers 8 inches in diameter—are scattered throughout the forecast area, each one often representing many square miles. In the Midwest the average gauge represents 250 square miles, although the U.S. average is one per 100 square miles. Measurements of rain collected in rain gauges are taken daily, either by hand or by electronic devices, and reported.

Since rain rarely falls evenly, this method of calculating rainfall might be likened to measuring the height of one meteorologist in each state and then saying that Missouri meteorologists are taller than Maryland meteorologists. Nevertheless, even a general idea of rainfall is valuable. A rain gauge *is* accurate for one place—your house, for example, or your garden or farm. You can make a rain gauge from any straight-sided, flat-bottomed container—a large juice can, for example. Ideally, it's set 3 feet above the ground and away from buildings, trees, or other interferences.

The meteorological year runs from July through June, so that winter statistics won't be split into two years, as would happen in a January–December year. Cumulative, year-to-date rainfall reported in most newspapers usually begins July 1, which is why late-summer precipitation statistics may appear surprisingly low.

ROMAN NUMERALS

Like Dracula, Roman numerals rise from their coffin in the dead of night and stamp themselves onto buildings, slip into books, number chapters, and force themselves into student outlines. Decoding Roman numerals isn't really difficult, though, and might even be an amusing pastime.

Here are the letters used to symbolize numbers in the Roman numeral system:

$$I = 1 \qquad C = 100$$
$$V = 5 \qquad D = 500$$
$$X = 10 \qquad M = 1,000$$
$$L = 50 \qquad \text{There is no zero.}$$

To read a number, simply add up the numbers from left to right (large numbers to small ones):

Roman Numbers	Arabic Numerals
II	$1 + 1 = 2$
III	$1 + 1 + 1 = 3$
VIII = V + III	$8 = 5 + 3$
XXVII = XX + V + II	$27 = 20 + 5 + 2$
LXXXVI = L + XXX + V + I	$86 = 50 + 30 + 5 + 1$
CCLI = CC + L + I	$251 = 200 + 50 + 1$
MMMCCCVI = MMM + CCC + V + I	$3306 = 3000 + 300 + 5 + 1$

The Romans, much like the rest of us, were always looking for an easier way to do things, and so after a while, subtraction was introduced to shorten the bulkier numbers. The subtracting involves only the numbers 4 and 9 wherever they occur, including 14 and 19, 24 and 29, etc., as well as 40 and 90, 400 and 900, and 4,000 (technically, Roman numerals can count only to 4,999). So you subtract smaller numerals that precede larger ones: instead of writing MMMMCCCCXXXXIIII ($4,000 + 400 + 40 + 4 = 4,444$), you can shorten it to MMMMCDXLIV ($4,000 + [500 - 100] + [50 - 10] + [5 - 1]$). Here are some easier examples:

$$\text{IV} \quad = \quad V - I = 4$$
$$\text{IX} \quad = \quad X - I = 9$$
$$\text{XIV} \quad = \quad X + (V - I) = 14$$
$$\text{XL} \quad = \quad L - X = 40$$
$$\text{XCIX} \quad = \quad (C - X) + (X - I) = 99$$

Adding and subtracting Roman numerals—much less multiplying and dividing them—is quite literally unthinkable. Hence, their demise.

RUBBER BANDS

Rubber bands are speedy, reusable, fun to shoot, and cheap. The National Office Products Association claims that if a rubber band saves one second of working time, it pays for itself. But wait—what do those strange size numbers mean? Rubber band sizes are confusing, because the size numbers are arbitrary, not being *of* anything, but representing a length-width combination. Furthermore, as you will see below, a larger number doesn't always mean a longer

ACTUAL RUBBER BAND SIZES

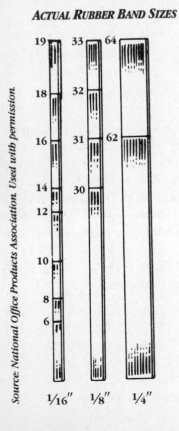

Source: National Office Products Association. Used with permission.

STANDARD RUBBER BAND SIZES IN INCHES		
Size	Length	Width
8	⅞	1/16
10	1¼	1/16
12	1¾	1/16
14	2	1/16
16	2½	1/16
18	3	1/16
19	3½	1/16
30	2	⅛
31	2½	⅛
32	3	⅛
33	3½	⅛
62	2½	¼
64	3½	¼
73	3	⅜
84	3½	½
105	5	⅝
107	7	⅝
54	Assortment	. . . .

rubber band. There is some method to the madness, however; with the exception of size 8, all the sizes in the same *decade,* such as teens, twenties, and so on, are the same width, and the widths do increase as the tens get higher; for example, a size in the teens is always 1/16 inch thick, while a size in the 30s is always 1/14 inch thick, etc. And the higher the number *within a decade,* the longer the rubber band.

SANDPAPER

Sandpaper isn't really sandpaper anymore—it's "coated abrasive." The original material was paper covered with grains of flint, which was inexpensive but didn't last long and made slow progress. Today, although flint is still available, other abrasives are more popular. Garnet is the hardest natural abrasive used in sandpaper, and it lasts five times longer than flint. Synthetic papers are even more popular—aluminum oxide is widely used as an all-around paper, and silicon carbide is the hardest of all the synthetics. There are other "abrasives," as they are called now, as well.

The old system for grading grits on abrasive papers was a 1 to 12 system, with the low numbers being coarse and the high number leading to finest. Today that system has been replaced with a 12 to 600 system, which also goes from low numbers for coarse to high for incredibly fine. The numbers refer to the size of the grit on the paper, *not* the number of grits per square inch, although the numbers probably do refer to the number of grits that could fit in a defined area. The amount of grit on sandpaper, however, can vary—"open coat" sandpaper has fewer grits, cutting faster and cleaning more easily than "closed coat" sandpaper, on which the grits are closer together.

There are usually 22 grades of abrasive paper, not 600 as the numbers seem to imply: 12, 16, 20, 24, 30, 36, 40, 50, 60, 80, 100, 120, 150, 180, 220, 240, 280, 320, 360, 400, 500, 600. The papers are also marked with an A, C, or D (there is no B) to indicate the weight of the backing—A is the lightest. As with many hardware items, the adjectives applied to the various grades are not always the same—one company's "fine" may be between 100 and 120, while another's may be as coarse as 80. The important thing is the number.

SANDPAPER GRADES

240–600	Extra fine
150–220	Very fine
100–120	Fine
60–80	Medium
30–50	Coarse

Which sandpaper do you use for what? It really depends on a number of factors, but these suggestions from several heavy users of sandpaper serve as a "rough" guide:

40 or 50	De-rusting metals Stripping housepaint Getting rid of a lot of wood fast for shaping
60	Coarse sanding for wood, shaping Stripping wood floors Grinding welding joints Rough metal work
80	General joint alignment in woodworking Rough metal finishing Stripping painted furniture Flattening board laminations
100	General medium woodworking Smoothing joints Autobody finishing
120	Final prepainting finish for furniture
220	Fine sanding for wood (higher is polishing)
220–400	Smoothing paints on wood or metal
400–600	Metal polishing Polishing glass edges

SCREWS AND BOLTS

Few things are as baffling to the uninitiated as the huge variety of threaded products—and their associated numbers—in a hardware screw and bolt department.

The first problem is deciding what's a screw and what's not. There is some energetic difference of opinion about this—one authority insists that a screw comes to a point and a bolt ends flat and requires a nut or hole with mating threads. Another, confronted with the nontapering machine screw, feels strongly that any threaded fastener is a screw until you put a nut on it, in which case, it's a bolt. The truth is, it doesn't really matter—screws and bolts are sized pretty much the same way.

Then there is the dazzling array of screws. If you want to buy a screw, you need to know the size, the material it's made of and/or the material you're screwing it into, and the kind of head. FH is a flat head screw; RH is a round head. There are also hex head, pan head, and Phillips, not to mention lag screws, sheet metal screws, square drive screws, and machine screws.

Then, there is the problem of the screw size. There are often three numbers sizing screws, usually written something like this: 4–40 × ½, sometimes written ¼₀ × ½.

The first number—the 4—is the *size.* The American Society of Mechanical Engineers (ASME), responsible for most screw and bolt sizing, uses one system for small screws and another for larger screws. Screws with a diameter under ¼ inch are commonly numbered from 000 to 12. The smaller the number, the smaller the diameter of the screw as measured just under the head. Screws ¼ inch in diameter or larger are usually sized by the fraction of an inch. So instead of a 0 or 6, you'll see a fraction, like ⅜ or ⁵⁄₁₆ or ½.

The second number, which usually comes after a hyphen or slash, is the *pitch*—the number of threads per inch.

SOME COMMON SIZE-PITCH
COMBINATIONS FOR SCREWS

2–56	8–32	¼–20
4–40	10–24	¼–28
6–32	10–32	¼–32

The number after the "×" is the length of the screw in inches. Size and thread pitch combinations, such as those above, come in various lengths. A size 4–40 × ½ is a size 4 screw with 40 threads to the inch and ½ inch long. But some boxes of screws are labeled with the length first, coming before the "×," but it's not hard to know whether the length is first or last—one of the numbers will correspond to the length you're looking at. You also have to contend with metric screws; metric measures are usually preceded with an "M," however, and are found in a special section.

There was a time when threaded fasteners were handmade and there was no standardization at all. Each nut had to be carefully paired with its bolt, for no other would fit it. As better lathes were developed for cutting the threads, the nuts and bolts from one shop would be interchangeable, but not with those from another shop. Coming up with standard screw, nut, and bolt sizes kept machinists, inventors, and governments busy for hundreds of years. It was about 1800 that David Wilkinson invented the screw-cutting lathe in America that could cut precision screws and bolts that matched. Today, in a process called thread-rolling, blanks are reformed using threaded dies. Nothing is cut away.

The standardization of screw threads was a longer battle, however, continuing between the United States and the British right into World War II, when most of the parts for war equipment from Great Britain and the United States were interchangeable—except the screws needed to put them together! The problem was that the British proposed the first standards for screw threads in 1841, which set the angle between the sides of the threads at 55 degrees. Then in 1864 the United States decided to set its own standard of 60 degrees, finding it stronger and easier to gauge than the British 55-degree angle. Each country held stubbornly to its own idea of the perfect screw until 1960, when the 60-degree angle was accepted by both, along with a standard of diameters and threads-per-inch for each.

It was about this time that the world was pushing metrics in every possible area of measurement. The International Organization for Standardization began work on a worldwide standard screw thread system in 1964 that would somehow bring the metric screw system and the United Thread System (the U.S.-British system) together. They are still working at it. For most purposes, metric threaded fasteners must still be purchased specifically for metric equipment.

SHIPS

Although most recreational watercraft are described in terms of length (feet), large commercial vessels have outgrown this description and today are described in terms of *tons*. "Tonnage" when applied to a ship can have either of two meanings—it can refer to the volume of available space or it can refer to weight—but a book or magazine describing a "20,000-ton ship" as often as not will fail to explain which kind of ton it's referring to. And there is a big difference.

GROSS TONNAGE *Gross tonnage,* also called *gross register tonnage,* or *gross weight,* is not weight at all. Gross tonnage is the total cubic capacity of a ship (all enclosed space, less a few spaces such as hatchways and others) as expressed in register tons, with 1 register ton equaling 100 cubic feet. Gross tonnage is a measure of volume, even if it is sometimes referred to as "weight": A ship of 20,000 gross tons has 2,000,000 cubic feet of enclosed space. Although most often applied to passenger ships, it's not unusual to see it used with other ships as well.

This use of tonnage to measure volume began back in the thirteenth century when merchant ships frequently carried wine in giant casks called *tuns*; a ship's carrying capacity came to be measured by the number of tuns she could carry. By the fifteenth century England had established a standard wine-filled tun of 250 gallons weighing 2,240 pounds. The 100-cubic-foot ton, which replaced it in the nineteenth century, remains the most common comparative measure of ship size. (*Net tonnage,* not often encountered in general sources, is actual carrying capacity—the gross tonnage less specified spaces unavailable for cargo, such as the engine room, crew's quarters, etc.)

How do gross tons translate into ship size? Columbus's *Santa Maria* is said to have been about 150 tons, the *Mayflower* 180 tons, the average clipper ship about 1,000 tons, with nineteenth-century sailing ships reaching up to 5,000 tons. The British steamship *Great Eastern,* launched in 1859, was almost 19,000 gross register tons and for nearly fifty years was the biggest ship in the world. But passenger ships got bigger and bigger—the *Titanic,* which sank on its 1912 maiden voyage, was 46,000 tons—culminating in the 1930s with the three biggest: the *Queen Mary* at 81,000 gross tons and the *Queen Elizabeth* and the French *Normandie* at 83,000 gross tons each. These huge ships proved less efficient than smaller passenger ships, however; today the average cruise ship is about 20,000 gross register tons.

DISPLACEMENT TONNAGE Displacement tonnage does refer to weight, but not the actual weight of the ship. Displacement tonnage is the weight of the water *displaced* by the ship. Because physically weighing a ship would be awkward, if not impossible, the submerged area of the ship is calculated in cubic feet. A floating object always displaces an amount of water equal to the area under the water, so the number of cubic feet of water displaced will be the same as the number of submerged cubic feet of ship. Multiply this figure by the weight of a cubic foot of water and you have the weight of water displaced. This figure is expressed using the long ton of 2,240 pounds (the 2,000-pound short ton is rarely used in the shipping business).

Although this may seem an odd way to weigh something, the system works well enough. Sometimes that's not enough to know, however; you may also need to know which type of displacement weight is being referred to.

Loaded displacement tonnage is the weight of the water displaced when a ship is carrying its normal fuel, crew, and cargo load. When a ship is described in "displacement tons," this usually refers to loaded displacement tons. Most naval ships are described in displacement tons: The full load displacement weight of an aircraft carrier can be about 80,000 tons; a battleship up to 59,000 tons; a destroyer 4,000 to 10,000 tons.

Light displacement tonnage is the weight of the water displaced by the unloaded ship.

Dead weight tonnage is the difference between the first two, which amounts to the weight of cargo, stores, water, fuel, crew, and passengers (in terms of the weight of water displaced) that the ship can carry. Freighters and tankers are usually described in terms of dead weight tons: A typical cargo vessel of 9,200 gross tons (cubic foot capacity) may average about 21,000 tons displacement (meaning loaded displacement) and 13,500 tons deadweight. A 300,000-ton oil tanker is no longer unusual—tankers are now reaching almost 500,000 dead weight tons, with a million-ton tanker in the works!

SHOES

Shoe sizes in this country really make no sense at all, and maybe the reason for this is that our shoe sizes have been based on barleycorns since the seventh century. This system became more or less official when, in 1324, King Edward II decreed the inch to be the length of three barleycorns. Shoemakers discovered that the longest normal foot around was as long as 39 barleycorns in a row, which was exactly 13 inches. Thus 13 became the largest shoe size, with full sizes equaling one barleycorn ($\frac{1}{3}$ inch) each. So a size 12 is 38 barleycorns, a size 11 is 37 barleycorns, a size 10 is 36 barleycorns, and so on. This system found its way to America, and in 1880 Edwin B. Simpson of New York introduced the first universally adopted system for shoe sizing, which remains to this day, adjusted only slightly: The insole measurement of American shoes has decreased by $\frac{1}{12}$ inch and half sizes are available, equaling half a barleycorn ($\frac{1}{6}$ inch). Today, of course, size 13 is no longer the biggest size.

Unfortunately, American women's shoe sizes do not correspond to men's—a man's size 8 is about a woman's size 9½. Perhaps Mr. Simpson introduced this confusion. Children's shoe sizes are also odd, because the width of the hand (about 4 inches) was used to determine the smallest size, and the length of the hand (about 9 inches) the largest size. So children's sizes run from 0 (4 inches) to 13 (9 inches).

European shoe sizes are understandably unrelated to this complicated system. Their unit is still based on thirds, though: Each European size equals $\frac{2}{3}$ of a centimeter, with the biggest size usually a 50 (the equivalent of an American 14). Most of the world uses this metric shoe-sizing system, which doesn't require half sizes and applies to both male and female feet.

A new metric shoe-sizing system called Mondopoint has been introduced by SATRA, a British shoe research organization, which would solve the shoe confusion worldwide. However, although most countries say the shoe fits, Americans are not ready to give their comfortable barleycorns the boot.

SNOW

If you've just lived through a blizzard which meteorologists report dropped 16 inches of snow on your sidewalk, just how much actual precipitation is that? Snow, in its melted form, is figured into the annual precipitation figures. The rule of thumb for estimating this is that 10 inches of snow equals 1 inch of water. But anyone who has shoveled a few sidewalks knows that there's snow and there's snow—ten inches of light fluffy stuff is less heavy—and contains less water—than the wet, backbreaking kind. So measuring snow depth won't work for the statistical record.

Meteorologists solve this problem by collecting snow in rain gauges—straight-sided 8-inch containers—and in snow gauges, which are slightly larger. These measure not only the snowfall but the amount of water generated when it melts. In mountainous areas where the snowpack (deep layers of accumulated snowfalls) is an important source of water, the snow depth is not only measured, but core samples are taken to see what kind of snow is involved. Snow that melts or is rained on does not always run off immediately—the water can filter to the bottom and freeze into ice, as much as doubling the snow/water ratio. This kind of information is important not just to skiers, but to water- and flood-control agencies.

SOCIAL SECURITY NUMBERS

If you've ever wondered whether your Social Security number gives away secrets to people in the know, you can stop worrying.

WHERE SOCIAL SECURITY NUMBERS ARE ASSIGNED

001–003	New Hampshire	468–477	Minnesota
004–007	Maine	478–485	Iowa
008–009	Vermont	486–500	Missouri
010–034	Massachusetts	501–502	North Dakota
035–039	Rhode Island	503–504	South Dakota
040–049	Connecticut	505–508	Nebraska
050–134	New York	509–515	Kansas
135–158	New Jersey	516–517	Montana
159–211	Pennsylvania	518–519	Idaho
212–220	Maryland	520	Wyoming
221–222	Delaware	521–524	Colorado
223–231	Virginia	525–585	New Mexico
232–236	West Virginia	526–527 and 600–601	Arizona
232, 237–246	North Carolina		
247–251	South Carolina	528–529	Utah
252–260	Georgia	530	Nevada
261–267 and 589–595	Florida	531–539	Washington
		540–544	Oregon
268–302	Ohio	545–573 and 602–626	California
303–317	Indiana		
318–361	Illinois	574	Alaska
362–386	Michigan	575–576	Hawaii
387–399	Wisconsin	577–579	District of Columbia
400–407	Kentucky		
408–415	Tennessee	580	Virgin Islands
416–424	Alabama	580–584 and 596–599	Puerto Rico
425–428 and 587–588	Mississippi		
		586	Guam
429–432	Arkansas	586	American Samoa
433–439	Louisiana	586	Philippine Islands
440–448	Oklahoma	*700–728	Railroad Board
449–467	Texas		

*700–728 was reserved for railroad employees. New number issuance in 700-series was discontinued in 1963.

Note: Some numbers are shown more than once because either they have been transferred from one state to another or they have been divided for use among certain geographic locations.

Source: Social Security Administration, U.S. Department of Health and Human Services.

Although you are the only person holding your particular Social Security number, the numbers reveal little about you.

Social Security numbers are divided into three parts: The first three digits currently represent the number holder's state of residence at the time the number was issued. Before 1973 the number simply established that the card was issued in one of the Social Security offices in that state. Today the first three digits are determined by—but are not the same as—the ZIP code of the mailing address shown on the application for a Social Security number. They are assigned by the central office.

The second group of numbers is the Social Security Administration's code number for the year the card was issued, beginning with 01 and proceeding in a set pattern—not necessarily in sequence—from there. Number 385-42-0000, for example, does not mean that the card was issued in 1942—42 has been issued in different years by different states, depending on how quickly each of their three-digit state codes worked through the sequence. A very low number, such as -01- or -02-, probably means that the card was issued about fifty years ago, when numbers were first assigned. However, that is not always true. A new batch of 600 numbers has recently been assigned to California, so 619-01-0000 could have been issued in 1988, the first year that California began using 600 numbers. A very high number, such as 459-95-0000, probably is fairly recent.

The third group—four digits—is, taken with the others, your personal number, assigned at random.

SOCKS

You have it from the sock authority of the United States of America, the president of the National Association of Hosiery Manufacturers (NAHM)—sock sizes correspond approximately to the actual length of your foot, in inches. If your foot is 10 inches long, your sock size is probably also 10. How did such a sensible system begin? One can only assume that it was first applied to men's socks, men's clothing being the only area of apparel in which sizes are based on real measurements.

Before the 1940s socks were made of natural fibers such as cotton, wool, and silk. Natural fibers are rigid, so socks had to be sized to the last half-inch. Each pair of socks had a specific size, such as 9, 9½, 10, 10½, and on up. A size 9 sock fit a foot measuring 9 inches, and so on.

Then, in 1939, nylon, the first synthetic fiber, was invented, and by the early 1950s yarn producers had figured out how to make it stretch and, much like a telephone cord, snap back to its original shape. This stretchiness was God's gift to the sock people, who could now put a wide range of foot sizes into the same-size sock. Not only did this save lots of money; the sock actually fit better, clinging to whatever foot it surrounded. So today, if your foot is 10 inches, you wear a size 9–11 sock. Fortunately, you don't really need to know how long your foot is; you probably have known your sock size for quite a while.

However, socks for little feet—infants' and children's—correspond more to their shoe size than the actual foot measurement; the hosiery industry is predicting that all sock sizes someday may be designed to approximate shoe sizes more closely.

SODIUM (SALT)

When you read the sodium content on a food label, how do you know whether or not to be shocked? How much sodium should you—or shouldn't you—have? The recommended amount for adults is under 3,300 mg. per day—the 6,000 to 12,000 mg. consumed daily by the average American is too much. But how much *is* that?

Consider this: 1 teaspoon of salt contains about 2,000 mg. of sodium, meaning that 1½ teaspoons (3,000 mg.) per day is about right. If you average 8,000 mg. daily, you are consuming 1⅓ tablespoons of salt daily, the equivalent of 2 *cups* of salt every month.

Unfortunately, you can't always tell the amount of sodium in food by taste. A regular order of McDonald's French fries (109 mg.) contains less sodium than a glass of milk or a slice of bread (about 140 mg. each), but according to their 1986 published figures, their cherry pie may contain almost four times that much (about 398 mg.). About one-third of the sodium you eat occurs naturally in fresh foods; the rest is added during processing, cooking, and at the table. Processed and fast foods are some of the worst sodium offenders. The easiest way to cut your sodium intake is to avoid them and go light on the shaker. To gauge how heavy your hand is on the shaker, sprinkle a piece of paper as you normally do a plate of food; then measure the salt in a measuring spoon.

SOME SODIUM COMPARISONS

Low-Sodium Foods	mg.	High-Sodium Foods	mg.
¼ lb. hamburger and bun	222	McDonald's Quarter Pounder	718
¼ lb. (Cheddar) cheeseburger & bun	544	McDonald's Quarter Pounder with cheese	1,220
1 fillet halibut, broiled in butter	168	McDonald's Fillet-o-Fish	800
Homemade chili with ground beef & canned low-salt tomatoes & kidney beans	370	Canned chili con carne with beans	1,354
7 oz. chunk tuna, water pack	75	7 oz. chunk tuna, oil pack	1,584
Homemade chicken or beef broth (no salt added)	50	Bouillon cube (just 1!)	960
		Many canned soups	1,000–2,000
1 cup boiled carrots	50	1 cup canned beef broth	1,597
1 cup homemade beef and vegetable stew, with lean chuck	91	1 cup canned carrots	581
		1 cup canned stew	1,007

Source: McDonald's figures from McDonald's Food: The Facts . . . , *McDonald's Corporation, 1986. All other figures from* Nutritive Value of American Foods, *by Catherine F. Adams, United States Department of Agriculture.*

SOIL, GARDEN

Even if you're an avid gardener who reads about soil pH with some frequency, you may still be among the many who are confused about the subject. In fact, if you're serious about gardening, you probably had your soil tested. That's often a good idea—many garden soils are too acid, meaning that the pH is too low (see **pH,** page 132). As a rule of thumb, the more rain you get, the more acidic your soil is likely to be; soils in a dry climate tend to be alkaline, containing chalk, lime, or other such substances.

Why should you be concerned about pH? The pH of soil affects the plants' ability to use the nutrients from organic matter in the soil as well as the fertilizer you may add. If the pH is too low or too high, a nutrient may not remain soluble long enough to reach the roots. The pH also affects the number of microrganisms in the soil, which help change organic nitrogen to a form that the plants can use.

The best pH for most gardens (flower and vegetable) is 6.5—a level at which nearly any plant except "acid-loving" plants can grow. The chart below offers general guidelines to the pH level most plants prefer, but there are always exceptions to every rule. (The pH of most ready-mixed soils you buy at the garden shop is printed right on the bag.) Public libraries and bookstores can provide more detailed information.

A GARDEN OF pHs

Range or Average pH	Example
2.7 to 3.4	Swamp peat
4.0 to 8.0	Survival range for most plants
4.5 to 5.5	Rhododendrons and azaleas
4.5 to 4.7	Average soil
5.5 to 6.5	Woodland ferns
5.5 to 7.2	Reasonable pH for most plants
Under 6.0	Considered to be "acid soil"
6.0 to 6.8	Most annuals
6.0 to 7.0	Best pH for most plants
6.0 to 7.0	Best for vegetables*
6.0 to 7.0	Best for lawns
6.5 to 7.0	Most perennials and vines
9.0 to 11.0	Desert soils
9.4	Lime (calcium carbonate)

*Some vegetables that will grow in 5.5 soil (as well as in the optimum 6.0 to 7.0 soil) are cucumbers, eggplants, onions, peppers, potatoes, pumpkins, rhubarb, sweet potatoes, tomatoes, and turnips. Potatoes, both white and sweet, are especially fond of acid soil.

SOIL TESTING If you're curious about the pH of your garden soil, have it tested. It rarely costs very much and serves a dual purpose, for the experts who do the testing are often the same ones who will tell you what to do if your soil pH is too low or too high. How to find the testing agency? In California, the only state to date without a public agency for this, look under "Laboratories" in the Yellow Pages. In other states call your county agricultural agent; if you can't find this in your telephone book, write to the Cooperative Extension Service of the appropriate state university:

COOPERATIVE EXTENSION OFFICES

Alabama	Auburn U., Duncan Hall, Auburn 36849
Alaska	U. of Alaska, Fairbanks 99775
Arizona	U. of Arizona, Tucson 85721
Arkansas	U. of Arkansas, Box 391, Little Rock 72203
California	U. of California, 2120 University Ave., Berkeley 94720
Colorado	Colorado State U., Fort Collins 80523
Connecticut	U. of Connecticut, Storrs 06268
Delaware	U. of Delaware, Newark 19717
District of Columbia	U. of District of Columbia, 4200 Connecticut Ave. NW, Washington, D.C. 20008
Florida	U. of Florida, McCarty Hall, Gainesville 32611
Georgia	U. of Georgia, Athens 30602
Hawaii	U. of Hawaii at Manoa, Honolulu 96822
Idaho	U. of Idaho, Moscow 83843
Illinois	U. of Illinois, 122 Mumford Hall, Urbana 61801
Indiana	Purdue U., Lafayette 47906
Iowa	Iowa State U., Ames 50011
Kansas	Kansas State U., Manhattan 66506
Kentucky	U. of Kentucky, Lexington 40546
Louisiana	Louisiana State U., Baton Rouge 70803
Maine	U. of Maine, Orono 04469
Maryland	U. of Maryland, College Park 20742
Massachusetts	U. of Massachusetts, Amherst 01003
Michigan	Michigan State U., East Lansing 48824
Minnesota	U. of Minnesota, St. Paul 55108
Mississippi	Mississippi State U., Mississippi State 39762

Missouri	U. of Missouri, Columbia 65211
Montana	Montana State U., Bozeman 59717
Nebraska	U. of Nebraska, Lincoln 68583
Nevada	U. of Nevada, Reno 89557
New Hampshire	U. of New Hampshire, Durham 03824
New Jersey	Rutgers U., Cook College, New Brunswick 08903
New Mexico	New Mexico State U., Las Cruces 88003
New York	Cornell University, Roberts Hall, Ithaca 14853
North Carolina	North Carolina State U., Raleigh 27695
North Dakota	North Dakota State U., Fargo 58105
Ohio	Ohio State U., 2120 Fyffe Rd., Columbus 43210
Oklahoma	Oklahoma State U., Stillwater 74078
Oregon	Oregon State U., Corvallis 97331
Pennsylvania	Pennsylvania State University, University Park 16802
Rhode Island	U. of Rhode Island, Kingston 02881
South Carolina	Clemson U., Clemson 29634
South Dakota	South Dakota State U., Brookings 57007
Tennessee	U. of Tennessee, Box 1071, Knoxville 37901
Texas	Texas A&M U., College Station 77843
Utah	Utah State U., Logan 84322
Vermont	U. of Vermont, Burlington 05405
Virginia	Virginia Polytechnic Institute, Blacksburg 24061
Washington	Washington State U., Pullman 99164
West Virginia	West Virginia U., Morgantown 26506
Wisconsin	U. of Wisconsin, 432 N. Lake St., Madison 53706
Wyoming	U. of Wyoming, Box 3354 U. Station, Laramie 82071
Puerto Rico	U. of Puerto Rico, Mayaguez 00708

Once you get your pH results, be sure to get expert advice before you try to correct your soil. Each type of soil (sandy, loam, and clay) requires a different treatment—the more clay in acid soil, the more lime will be needed to neutralize it. It can take years to correct the soil down to the root zone, so if your soil is quite a bit off, get it tested every year for a while to see how it's progressing.

S*OUND*

A sound is a vibration which travels in waves through the air or some other medium. But what does sound *sound* like and how is it measured? This is determined through pitch, loudness, and speed.

*P*ITCH   The pitch of a sound—how high or low it sounds—depends on the frequency of its waves, or how fast the waves pass a particular point. High-pitched tones are produced by quickly vibrating, high-frequency waves; low tones by more relaxed, low-frequency waves. This frequency is measured in *hertz* (see **Radio Waves**, page 149), the number of waves (or cycles) per second. The lowest note on a piano, for example, has a frequency of 27 Hz, while the highest note is measured at 4,000 Hz.

Humans with normal hearing can hear sounds even higher than that—up to 20,000 Hz. Sounds too low for human hearing (those below 20 Hz) are called infrasonic; those too high (over 20,000 Hz), ultrasonic. It's not unusual for animals to have infrasonic and/or ultrasonic hearing:

Humans	20–20,000 Hz
Dogs	15–50,000 Hz
Cats	60–65,000 Hz
Bats	1,000–120,000 Hz
Dolphins	150–150,000 Hz

*L*OUDNESS   Most of us have assumed that loudness is measured in *decibels.* Technically, this is not true—loudness is measured in *phons,* a combination of decibels and hertz.

The decibel is a level of a sound's energy, or power, or intensity. This energy can be measured in watts per square meter, but sound generates so little energy (the sound energy generated by a rock band might barely power a 100-watt light bulb) that these numbers are awkward—for example, 10 decibels equals .000000000001 watt per square meter. Decibels are decidedly easier to use.

The decibel is not an actual unit of measurement, but a comparison between one sound and another. In 1935–36, the U.S. government tested 5,000 people for the softest sound they could hear. This sound level became the reference point, or 0 decibels (db), representing minimum human hearing. Thus 0 decibels is not an absence of sound, but the softest sound audible to the human ear. Here's how some sounds measure in decibels:

DECIBELS OF SOME COMMON SOUNDS

Decibels	Effect on Human Hearing	Example
0	Hearing begins	
10	Just audible	Leaves rustling
30	Very quiet	Ticking clock at 1 yard, library, whisper
40		Quiet office, bedroom
50	Quiet	Light auto traffic
60	Intrusive	Air conditioning unit at 20 feet
70		Freeway traffic, noisy restaurant (conversation is difficult)
80	Annoying	Motorcycle, hair dryer, alarm clock
90	Very annoying	Heavy truck, city traffic, lawn mower
100		Inside a subway, auto horn, chain saw
110		Amplified rock music
115		Jet aircraft engine at 20 feet
120	Uncomfortable	Very close thunder clap
130	Damaging	Air raid siren
140–150	Painful	Jet taking off

What do those numbers actually mean? Is a motorcycle (at 80 decibels) only twice as loud as your bedroom (at 40 decibels)? Hardly. The decibel scale is logarithmic, similar to the Richter scale (see **Earthquakes,** page 58). Although to the human ear a 10-decibel increase doubles the loudness, *a 3-decibel increase (approximately) doubles the actual intensity of the sound.* In other words, if the noise made by one jet engine measures 115 db, two jet engines would measure 118 db. It works out like this:

THE DECIBEL SCALE

Decibels	Intensity (read "loudness" if you like)
0	minimum human hearing
10	10 times more than 0
20	100 times more than 0
30	1,000 times more than 0
40	10,000 times more than 0
50	100,000 times more than 0
60	1,000,000 times more than 0
70	10,000,000 times more than 0
80	100,000,000 times more than 0
90	1,000,000,000 times more than 0
100	10,000,000,000 times more than 0
110	100,000,000,000 times more than 0

SPEED Sound, unlike light, does not travel at a constant speed—how fast sound travels depends on what it's moving through. Although sound moves through air at about 750 miles per hour at sea level and 32 degrees Fahrenheit, it can zip through water at about 3,204 miles per hour, through brick at 8,114 miles per hour, and steel at 11,181 miles per hour.

Aircraft that fly faster than the speed of sound fly at *supersonic* speeds. *Mach 1* is the speed of sound, *Mach 2* is twice the speed of sound, and so on. The Mach numbers are used because the speed of sound depends on the altitude and temperature, and is found by dividing the speed of the aircraft by the speed of sound at that altitude. A plane flying 740 miles per hour at 40,000, where sound travels at 660 miles per hour, would be 740 divided by 660, or Mach 1.12.

STAPLES

In the 1800s, staples were cut from tin with a mallet. It wasn't until the early 1900s that staples were made of wire and preformed into their familiar shape. They were packed loose in a box—imagine trying to load up a stapler with individual staples. Today preformed wire staples are held with glue in handy strips of exactly 210 staples each for standard staplers and half strips of 105 for smaller staplers. These are usually packed 24 to a box, or just over 5,000 staples.

Standard office staples are usually made of round, galvanized wire .019 inch thick, are ½ inch wide, and have ¼ inch "legs." Flat staples for staple guns use heavier wire—staples using wire .024 inch thick or more are considered heavy duty. Both are sized by length, which refers to the length of the "legs," in fractions of an inch, from 5/22 inch to ½ inch for round staples, and from 11/64 inch to 9/16 inch for flat staples.

STEEL WOOL

All brands of steel wool are sized in much the same way, although the markings may be dissimilar. For example, #3/0 is the same as #000 (not #3). Sometimes the descriptions vary, too, but the smaller the number (or the more zeros there are) the finer the steel wool:

STEEL WOOL SIZES

#3	#2	#1	#0	#00 or #2/0	#000 or #3/0	#0000 or #4/0
coarse	medium coarse	medium	fine or medium fine	very fine or fine	extra fine	finest

STREET ADDRESSES

House numbers tend to be odd on one side of the street and even on the other. You can count on it. But even after years in a town, you may not know which side of the street to squint at in order to locate odd-numbered 735 Prospect Street. Why is that?

It's because you never know. It's easy to assume that a sensible system is at work because there are numbers involved and numbers imply order. There probably is a sensible system, but frequently several different systems apply in the same town. During the early growth of many cities and towns, small, widely separated areas often developed workable systems for themselves that years later clashed when the areas grew together and the systems bumped into each other. So while you might find the odd numbers on most streets in your town on the south or west side, they may occasionally pop up on the north and east sides, or vice versa.

Surprisingly, only in the last ten or twenty years have most cities and towns made a serious effort to make sense of a hundred, possibly two hundred years of individual and community whims and official fancies. This late response to the street name and number question is probably because, before computers became common, it was all just too intimidating. Now, however, sensible numbering and nonduplication of street names are crucial to prompt emergency help: Police, fire, and ambulance services are usually the first to be consulted by city planning commissions, and most 911 calls are responded to with the help of computerized address information. Planners usually work closely with the U.S. Postal Service as well.

SUNSCREEN LOTION

How many times have you stared at sunscreen lotions, puzzling over whether to buy one labeled 6 or one labeled 15? What do those numbers mean? The number on a sunscreen is called SPF (Sun Protection Factor), a standard established and recognized by the Food and Drug Administration. The SPF is a multiplier for the length of time you can withstand exposure to the sun longer than you normally would. If you usually burn after a half-hour exposure to the sun, you can lengthen your time before sunburning to three hours by using a number 6 sunscreen (½ hour times 6), or to 7½ hours by using a number 15 sunscreen.

Although at the moment the FDA recognizes SPF ratings only up to 15, the numbers are getting higher—sunscreens with an SPF as high as 30 can be found now. Many doctors feel that even small amounts of sun can harm your skin, and recommend using a sunscreen with an SPF of at least 15 during all outdoor activity. Skin cancers can take as long as twenty years or more to show up. Thanks to the thinning atmospheric ozone layer, more ultraviolet rays are getting through, and the occurrence of skin cancers is increasing.

TEMPERATURE

The notion of switching from familiar Fahrenheit readings to those of Celsius—or is it Centigrade?—is very disorienting. Exclaiming about a 38-degree day doesn't have the same impact as the Fahrenheit equivalent of over 100 degrees, and "below-zero" Fahrenheit weather makes zero Celsius look tepid by comparison. Why all these annoying temperature scales?

The first temperature scale, the one Americans love so well, has been in existence since 1714 and was devised by German physicist Gabriel Daniel Fahrenheit. To establish a scale, one needs a fixed cold point and a fixed hot point. Fahrenheit chose a mixture of table salt and ice—the coldest temperature he could devise and then thought to be the coldest possible—as his cold point, and, curiously, body temperature, which he established as 96 degrees, as his hot point, dividing the space between into 96 parts. (He was wrong about body temperature—it was later established at 98.6 degrees.) On Fahrenheit's scale, water froze at 32 degrees and boiled at 212 degrees.

Only twenty years later, Anders Celsius, a Swedish scientist, introduced another scale which he felt was more sensibly based on the freezing and boiling points of water and was also metrically sound. He set the freezing point of water at 0 degrees and its boiling point at 100 degrees, dividing the space between into a tidy 100 parts. (Celsius degrees are 1.8 times bigger than Fahrenheit degrees.) Countries using the metric system and scientists worldwide, who have always preferred metric measurements, adopted the Celsius scale without hesitation. Degrees were called centigrade until 1948, when the name was officially changed by an international conference to honor its inventor. ("Centigrade" is still commonly used in the United States, however.)

To translate Fahrenheit degrees to Celsius: Subtract 32 and divide by 1.8, or ($°F − 32) ÷ 1.8$. To go from Celsius degrees to Fahrenheit: Multiply by 1.8 and add 32, or ($°C × 1.8) + 32$. Although most thermometers today are marked with both scales, you still may not know how many degrees C you need for a nice spring day.

F AND C TEMPERATURES YOU LIVE BY

Degrees Fahrenheit	Approx. Degrees Centigrade (Celsius)	Conditions
212	100	Water boils
104	40	Heat wave conditions
98.6	37	Normal body temperature
98.6	37	Very hot day
86	30	Very warm day
68	20	Mild spring day
50	10	Warm winter day
32	0	Freezing point of water
32	0	Cold day
20	−7	Very cold day
0	−18	Extremely cold day

Even Celsius had its drawbacks, however. It seemed silly, certainly unscientific, to have "minus" degrees, since temperature is a measure of heat. In 1948 British physicist Lord Kelvin introduced a new scale which put 0 degrees at absolute zero—the coldest possible temperature, at which molecules stop moving and gases vanish—and used Celsius degrees to measure up from there. On the Kelvin scale, water freezes at 273.15 Kelvins (the preferred usage is "Kelvins" or "K," not "degrees Kelvin") and boils at 373.15 Kelvins—exactly 100 degrees higher. This makes it easy to go back and forth from Celsius (C) to Kelvin (K)—since 0 degrees C is 273.15 K, you add 273.15 to Celsius degrees to get Kelvins and subtract it from Kelvins to get Celsius. Kelvin's scale is not only accepted by the scientific community; it is the final authority. Celsius and Fahrenheit are now both defined in terms of Kelvins, much like the yard is defined by the meter.

TIDE TABLES

Even if you don't live near the coast, you probably know that the tide comes in and the tide goes out in a fairly predictable manner. You may even know that it does this, for the most part, twice a day. Possibly you've heard that each high tide happens about 12 hours and 25 minutes after the last one (a correct average but a dangerous assumption). And that's where you stop. You'd think that if there were two tides a day, they'd be a tidy 12 hours apart, occurring at the same time daily. It's that extra 25 minutes that confuses you.

You can blame this inconvenience on the moon. The rhythmic rise and fall of ocean waters—in addition to the sun's gravitational pull, the whirl of the earth, and the weather—are caused by the moon's gravitational pull on the earth. Unfortunately, the lunar day is not in sync with the 24-hour solar day—it's about 50 minutes longer—making tide prediction a scheduling nightmare. Furthermore, the tides are so affected by the coastline and other geographical factors that high tide in one bay may be different from that in the next. This is why it's best, if you're planning a coastal outing— fishing, swimming, tide-pooling, clamming, or whatever—not to make any assumptions about tides but to consult the tide tables.

Where to find them? Coastal newspapers publish the tide heights and times for the day, almanacs have them for the bigger stations, and most places that issue fishing licenses offer a whole year's tide tables for the local area in little booklets—often free. These tide tables publications are based on the National Oceanic and Atmospheric Administration's (NOAA) careful figures and predictions, which are in turn based on careful observations and comparisons with a sample nineteen-year period.

A warning, though: Officials at NOAA complain about the lack of accuracy in some of these booklets, which, though based on NOAA figures, are occasionally "full of typos and printing errors." For serious planning, you might venture to a library's reference section to consult *Tide Tables* (current year), *West (or East) Coast of North and South America,* issued by NOAA, U.S. Department of Commerce (R 525.6 in many libraries). Even these figures are estimates, however; tides can be influenced by weather even as distant as a storm off the coast of Japan.

Local tide table formats vary, but they all function the same way. The tables give the high and low tide times and heights for one main area—the example figures apply to the San Francisco Golden Gate Bridge—for one year. Only a few pages are then required to list the "corrections"—how those times and tide heights differ from the main tables—for the other nearby areas. Fortunately, these differences never change, so only one set of figures needs to be listed for each place. Look at these tables and corrections:

TIDES AT THE GOLDEN GATE BRIDGE

JULY 1987

513

		LOW TIDE			HIGH TIDE			
		AM	Ht.	PM	Ht.	AM	Ht.	PM	Ht.

		AM	Ht.	PM	Ht.	AM	Ht.	PM	Ht.
	Sunrise 5:40	-PDT-				Sunset 8:27			
1	W	9:25	0.0	10:09	2.9	2:36	4.9	4:58	4.7
2	Th	10:02	0.4	11:12	2.5	3:25	4.4	5:30	4.9
3	F	10:39	0.9	----	----	4:27	3.9	6:01	5.2
4	Sa	12:21	2.0	(11:25	1.5)	5:47	3.5	6:36	5.5
5	Su	1:16	1.3	12:17	2.1	7:28	3.4	7:15	5.8
	Sunrise 5:43	-PDT-				Sunset 8:26			
6	M	2:12	0.6	1:16	2.6	9:10	3.6	7:58	6.2
7	Tu	3:04	−0.1	2:15	3.0	10:27	4.0	8:47	6.5
8	W	3:53	−0.8	3:14	3.2	11:30	4.3	9:36	6.8
9	Th	4:42	−1.3	4:11	3.3	(12:18	4.6)	10:29	7.1
10	F	5:30	−1.7	5:05	3.2	(1:04	4.8)	11:18	7.2
	Sunrise 5:46	-PDT-				Sunset 8:25			
11	Sa	6:18	−1.8	6:01	3.0	----	----	1:49	5.0
12	Su	7:06	−1.8	6:56	2.8	12:10	7.1	2:31	5.1
13	M	7:52	−1.5	7:54	2.5	1:05	6.8	3:13	5.2
14	Tu	8:38	−1.0	9:00	2.2	1:57	6.2	3:54	5.4
15	W	9:22	−0.4	10:11	1.9	2:57	5.5	4:33	5.6
	Sunrise 5:49	-PDT-				Sunset 8:22			
16	Th	10:08	0.4	11:24	1.5	4:03	4.8	5:16	5.7
17	F	10:56	1.2	----	----	5:21	4.1	6:01	5.9
18	Sa	12:38	1.1	(11:49	2.0)	6:54	3.7	6:45	6.0
19	Su	1:45	0.7	12:51	2.6	8:36	3.7	7:31	6.1
20	M	2:45	0.3	1:53	3.0	10:00	4.0	8:19	6.1
	Sunrise 5:53	-PDT-				Sunset 8:19			
21	Tu	3:37	0.0	2:56	3.3	11:03	4.3	9:06	6.2
22	W	4:22	−0.2	3:48	3.4	11:53	4.6	9:49	6.2
23	Th	4:59	−0.3	4:34	3.4	(12:32	4.7)	10:32	6.2
24	F	5:36	−0.4	5:16	3.3	(1:07	4.7)	11:11	6.1
25	Sa	6:11	−0.5	5:55	3.2	(1:39	4.7)	11:49	6.0
	Sunrise 5:57	-PDT-				Sunset 8:16			
26	Su	6:43	−0.5	6:31	3.0	----	----	2:08	4.7
27	M	7:12	−0.4	7:09	2.9	12:25	5.8	2:36	4.7
28	Tu	7:41	−0.2	7:49	2.7	1:00	5.6	3:01	4.8
29	W	8:09	0.1	8:34	2.5	1:39	5.2	3:25	4.9
30	Th	8:40	0.5	9:27	2.2	2:17	4.8	3:51	5.1
	Sunrise 6:01	-PDT-				Sunset 8:11			
31	F	9:11	1.0	10:22	1.9	3:09	4.3	4:23	5.3

❭ 4   ○ 10   ◖ 17   ● 25

Copyright 1987. "Easy Read" Tide Book. Published by Wilkins Creative Printing, Atascadero, California.

SAN FRANCISCO COAST CORRECTION TABLES

	High		Low	
	Time	Change	Time	Change
Monterey Monterey Bay	−1:16	−0.5	−0:58	0.0
Santa Cruz	−1:19	−0.5	−1:04	0.0
Halfmoon Bay	−1:10	−0.2	−0:56	0.0
San Francisco Bar	−0:39	−0.1	−0:37	0.0
Alcatraz Island	+0:10	+0.1	+0:12	0.0
San Francisco, North Point	+0:15	+0.1	+0:22	0.0
Oakland Pier	+0:29	+0.3	+0:42	0.0
Alameda	+0:35	+0.7	+0:42	0.0
Oakland Harbor, Grove Street ..	+0:29	+0.5	+0:36	0.0
Potrero Point	−0:29	+0.6	+0:40	0.0
Point Avisadero, Hunters Point ..	+0:30	+0.9	+0:43	0.0
San Mateo Bridge	+0:39	+1.9	+1:14	+0.1
Redwood Creek Entrance (inside)	+1:02	+3.3	+1:32	+0.1
Dumbarton Highway Bridge	+0:48	+2.7	+1:27	+0.1
Alviso (bridge), Alviso Slough ..	+1:20	+3.3	+2:18	+0.1
Sausalito	+0:09	−0.2	+0:13	0.0
Angel Island (east side)	+0:22	0.0	+0:33	0.0
Richmond	+0:21	+0.1	+0:29	0.0
Point Richmond	+0:36	+0.2	+0:40	0.0
Point Orient	+0:47	+0.1	+0:54	0.0
Point San Quentin	+0:46	0.0	+0:58	0.0
McNear	−1:05	+0.1	+1:07	0.0
Pinole Point	+1.19	+0.4	+1:32	0.0
Hercules	+1:26	+0.4	+1:53	0.0
Petaluma River Entrance	+1:13	+0.5	+2:15	0.0
Selby	+1:25	+0.7	+1:58	0.0
Mare Island Strait Entrance	+1:41	+0.2	+2:03	−0.1
Napa, Napa River	+2:12	+1.5	+2:46	0.0
Crockett	+1:53	+0.2	+2:15	−0.1
Benicia, Army Point	+1:57	+0.3	+2:23	0.0
Port Chicago	+2:30	−0.2	+3:14	−0.1
Pittsburg, New York Slough	+3:23	−1.1	+4.14	−0.4
Point Buckler	+2:36	−0.2	+3:18	−0.3
Suisun Slough Entrance	+2:45	−0.3	+3:27	−0.3
Antioch	+3:56	*0.74	+4:42	*0.55
Three Mile Slough Entrance	+4:50	*0.61	+5:48	*0.45
Prisoners Point	+5:49	*0.61	+6:39	*0.45
Wards Island	+6:08	*0.63	+7:02	*0.45
Black Slough Landing	+6:24	*0.67	+7:25	*0.45
Stockton	+6:42	*0.71	+7:47	*0.45
Sacramento	+7:30	*0.51	+9:28	*0.27
Bolinas Bay	−0:29	0.0	−0:17	0.0
Point Reyes	−1:05	0.0	−0:37	+0.1
Tomales Bay Entrance	−0:16	*0.88	+0:10	*0.91
Bodega Harbor Entrance	−0:42	−0.1	−0:22	+0.1
Point Arena	−0:46	0.0	−0:27	0.0
Little River	−0:35	0.0	−0:25	0.0
Fort Bragg	−0:35	+0.1	−0:26	0.0
Shelter Cove	−0:43	+0.3	−0:23	+0.1

*Ratio - Multiply heights by this ratio and then apply the accompanying correction.

Local booklets like these usually account for daylight saving time, but always check—NOAA does not. Also, tide tables use decimal feet, dividing feet into tenths instead of inches. Now, let's find out the time and height of low tide in Half Moon Bay on the morning of July 16, 1987:

1. Check the San Francisco tables, which tell you low tide will be at 10:08 A.M. PDT (Pacific Daylight Time) and will be 0.4 foot. (How high is this? See below.)
2. Find Half Moon Bay in the correction table. These figures tell you that the low tide will be 56 minutes earlier (−0:56), or 9:12 A.M., and that there will be no change in the water height (0.0), so it remains 0.4.

That's all there is to it. Use the corrections to add or subtract for time and height (sometimes you have to multiply for water heights).

Some questions arise from tide tables, however. For example, does that 0.4 figure mean that the water under the Golden Gate Bridge is less than half a foot high at low tide? And what about those minus (−) figures in the main tables—what are they lower *than*? If you assume sea level, you're wrong. Not only is sea level different from place to place, but due to melting glaciers, it's rising. Because it's on the West Coast, the 0.4 foot figure tells you how much higher the water is than the *mean lower low tide* (which is simply the average of the lower of the two daily low tides over a certain 19-year period). On the East Coast, water heights are based on the *mean low tide* (the average of *all* the low tides). The tide tables do not tell you the actual depth of the water, only the differences from the base.

You may have noticed that few of the tides in the above samples are exactly 12 hours and 25 minutes apart—in fact, the low tides for July 27 are less than 12 hours apart and high tides are more than 14 hours apart. Because there are other strong influences on tide times, it could be dangerous to try to time the next high or low tide by adding 12 hours and 25 minutes to the last one—you don't want to be trapped on rocks or a rugged beach during a swift incoming tide.

If you're visiting the coast and don't want to sound like an inlander, here are a few terms for you:

Ebb current—the movement of water away from the shore.

Flood current—the movement of water toward the shore.

Neap tide—a slow tide, in which the difference between high and low water is less than usual.

Spring tide—a fast tide, in which the difference between high and low water is more extreme than usual. (This happens when the moon is extreme—new or full— not just in the spring.)

TIME

Asking someone for the time sounds like a simple request, and it is, usually. But if you start digging around into who actually determines what time it is, the answers—there's more than one—get complicated. Telling time accurately—and keeping the world on the same time—has stumped early time tellers as well as modern-day scientists. If you start to ask what time it is, you can run into apparent solar time, mean solar time, sidereal time, astronomical time, atomic time, Universal Time, U.S. Standard Time, and Greenwich Mean Time, to mention just a few. If you set a clock for each of these times, the settings will not all be exactly the same.

Apparent solar time is sundial time, noon determined by the moment when the sun is highest (it's rarely exactly overhead). This means that noon in the next town east or west will be several minutes earlier or later than noon in your town, so that the time is not the same in even nearby locations. Still, apparent solar time was used in this country until the railroads came along (see **Time Zones,** page 184).

Mean solar time bases time on the length of an *average day*. Because the earth does not spin at a uniform speed, day lengths in a year are not exactly the same.

Sidereal time tells time by the position of the stars, also called *astronomical time.* Sidereal time is tracked by a telescope—in our case, the Photographic Zenith Tube at the U.S. Naval Observatory in Washington, D.C.—in which the same stars appear at the same time every day. The mean sidereal day is 23 hours, 56 minutes, 4.09 seconds—the sun rises nearly 4 minutes later every day when measured by the stars. Because the earth's rotation can vary in speed, sidereal time is not perfectly uniform.

Atomic time is U.S. Bureau of Standards time, kept by atomic clocks (see **Time Units,** page 183). Atomic clocks, though incredibly accurate, are based on the behavior of cesium atoms, not on the behavior of the earth or the heavens; a second must be added to the end of every other year or so to keep atomic time in sync with the stars.

182 *TIME*

Universal Time (UT) is also known as *Greenwich Mean Time.* In 1972, key international groups concerned with time decided to keep laboratory clocks on atomic time, adding or subtracting "leap seconds" to the last minute of the year when needed and reporting to the Bureau Internationale de l'Heure (BIH) in Paris, which is the last word in establishing *Universal Time,* also called *BIH Time.* Today the BIH fields information from all over the world via a U.S. satellite and other channels, which it processes to help standardize time throughout the world. The whole world is not on exactly the same time. Even though a second can be timed to the billionth, getting everyone on the same nanosecond everywhere remains a challenge. (See also **Greenwich Mean Time,** page 89, **Military Time,** page 123, and **Time Zones,** page 184.)

Coordinated Universal Time (UTC) is Universal Time co-ordinated with the National Bureau of Standard's atomic clocks and the U.S. Naval Observatory's sidereal time. It is this time that is broadcast on the National Bureau of Standard's radio station WWV. If you call WWV—(303) 499-7111—you will get the exact time in Greenwich, England, using the 24-hour time system: The seconds are ticked off and the exact time is given every minute. (The military, by the way, relies on a different time signal, broadcast by the U.S. Naval Observatory.)

U.S. Standard Time is based on *Coordinated Universal Time (UTC):* Universal Time is simply translated into the correct time for your U.S. Standard Time Zone. (See **Time Zones,** page 184.) If you call "Time"—it's in your phone book—this is the time you'll get. It's usually based on WWV broadcasts.

Daylight saving time is U.S. Standard Time plus 1 hour, first introduced during World War I in 1916—if everyone gets up an hour earlier in spring and summer, when the sun rises earlier and sets later, much energy used for nighttime lighting is saved. Because adopting daylight saving time was up to local communities, however, confusion reigned—some used it; some didn't. As of 1987, all states adopting daylight saving time must begin it on the first Sunday in April and end on the last Sunday in October. A few states—like Arizona, Hawaii, and parts of Indiana—still don't use it.

TIME UNITS

Do you know how long a second is? You might say "one-sixtieth of a minute." Or maybe you'd push it further: "There are 60 seconds in a minute, 60 minutes in an hour, and 24 hours in a day, making 86,400 seconds in 24 hours ($60 \times 60 \times 24 = 86,400$). Therefore a second is $\frac{1}{86,400}$ of a day."

This perfectly sensible answer was the accepted measure of the second until 1820, when a committee of French scientists pointed out that the length of one day may vary a few thousandths of a second from another. So they measured a year of days and defined the second as $\frac{1}{86,400}$ of an *average* day, not just any day.

Then in 1956, scientists noticed that the earth's rotation is slowing down a bit—each day is about 20 microseconds longer than the corresponding day in the prior year—adding up to 7.3 milliseconds per year. Strangely enough, while the day is getting longer, the year is getting shorter—the earth's orbital period is 5.3 milliseconds less each year. In order to account for this, scientists decided to base the second on an average year. The official second—called the *ephemeris* second—became $\frac{1}{31,556,925.9747}$ of the year 1900, which was chosen as the *mean solar year*. But it's tough to remeasure the year 1900 in the laboratory. Furthermore, despite the impressive number of digits, the ephemeris second could be divided only into millionths, which was just not precise enough.

Since 1967 the second has been based on the absolutely predictable behavior of the cesium atom. In vastly oversimplified terms, the outer electron in the cesium atom flips when exposed to a radio signal of exactly 9,192,631,770 Hz. The quartz oscillator producing this signal is automatically adjusted to keep those electrons jumping, and its frequency, divided by 9,192,631,770 is defined as one second. This definition will probably remain for a while, as scientists can now split seconds into completely accurate microseconds (millionths of a second), nanoseconds (billionths), picoseconds (trillionths), and beyond. (To hear an atomic second tick by, call WWV—[303] 499-7111.)

The day is no longer our basic unit of time, divided into smaller units, nor is the year. Instead the tiny second, controlled by the U.S.

Bureau of Standards, is multiplied to define our days and years. We now live on atomic time, kept by atomic cesium clocks that measure time in nanoseconds. Even this isn't perfect, though. Now and then a second has to be added at the end of a year to keep us in sync with the stars—at the end of December 31, 1987, the fourteenth "leap-second" since 1972 was added.

TIME ZONES

Time zones are essential to keeping the world sunny-side up and making today's complex airline schedules possible. Before the late 1800s, however, no one gave much thought to schedules. Each town ran on sundial time, considering noon to be the moment when the sun was (more or less) directly overhead. Even a good horse took so long to get from one town to another that exact time hardly mattered. It was not essential for time to be precise until the railroads began traveling across the country. In part, standardized time and time zones resulted when people became regularly inconvenienced by the erratic time clock—there's hardly a more irate person than the one who's just missed his chosen mode of travel. The railroads *tried* to make schedules, but the railroad clocks had to be changed around so much that the schedules weren't very efficient—Michigan and Illinois each had 27 different local times, Indiana had 23, Wisconsin had 38, just for starters.

Finally, in 1883, the railroad companies adopted a time zone system that had been suggested by a teacher, Charles Gowd, which divided the United States into four time zones: Eastern, Central, Mountain, and Pacific. The width of each zone was the distance it took the sun to travel one hour, or 15 degrees between longitude lines that run from pole to pole (see **Latitude and Longitude,** page 107). After quite a bit of controversy the country adopted the railroad system, although the government didn't make it official until 1918. Other countries were immediately interested, however. In 1884 an international conference in Washington, D.C., divided the globe into twenty-four time zones, using the zero degree longitude line that runs through Greenwich, England, as the base line, and the 180 degree longitude line as the International Date Line.

STANDARD TIME ZONES OF THE WORLD

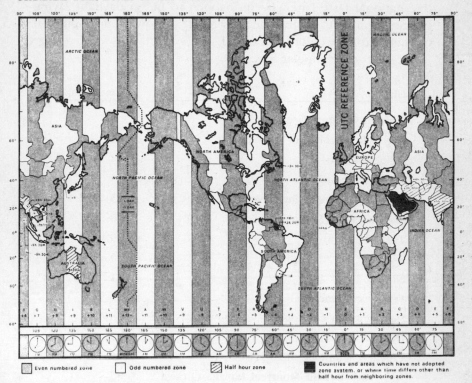

Source: From Sundials to Atomic Clocks *by James Jespersen and Jane Fitz-Randolph,* *U.S. Bureau of Standards.*

Today the time zone lines zig and zag to avoid splitting towns and smaller countries into different zones and other awkwardness. Zones to the west of Greenwich base line are numbered +1 to +11, each an hour earlier. Zones to the east are numbered −1 to −11, each an hour later. The line directly opposite the base line, on the 180 degree meridian, called the International Date Line, is +12. This makes the United States time zones +5 for Eastern Standard Time, +6 for Central Standard Time, +7 for Mountain Standard Time, and +8 for Pacific Standard Time.

The sanest way to establish the time at a foreign locale is to find out its time zone number and your zone number. If the foreign zone number is smaller than yours (−9 is smaller than +5, for example), figure it out going eastward. If it's bigger, count hours going west. This is to save you the headache of coping with the

International Date Line. (If you happen to be on daylight saving time, remember to subtract an hour from your own time to put you on Standard Time before you start.)

Here are some popular times:

City	Time Zone
London, Glasgow, Greenwich	0
Berlin, Rome, Madrid, Vienna, Oslo	−1
Istanbul, Cairo	−2
Moscow, Leningrad	−3
U.S.S.R. (encompassing eleven time zones!)	−3 to +11
Manila, Saigon, Shanghai	−8
Tokyo	−9
Sydney	−10
Nome	+11
Fairbanks, Honolulu	+10
San Francisco, Los Angeles, Vancouver	+8
Denver, Edmonton	+7
Chicago, Winnipeg, Mexico City	+6
New York, Detroit, Montreal, Havana, Ottawa	+5
Santiago, San Juan, St. Thomas	+4
Buenos Aires, Rio de Janeiro	+3

Try it. You are in New York (Zone +5) and it is 11:00 A.M. on Sunday. You want to know what time it is in Tokyo (Zone −9). You have to go 5 zones east to reach Greenwich and 9 more east to reach Tokyo, totaling 14 hours. Remember, as you go east, you are heading toward tomorrow, which is traveling westward, so when you pass midnight, you are in the next day, in this case, Monday. To make things easy, add 12 hours to take you to 11:00 P.M., and two more, passing midnight to make it 1 A.M. on Monday morning.

For a time zone numbered higher than yours (stay in Zone +5 for a moment), e.g., Honolulu (Zone +10), move westward: subtract the difference in the times (5 hours) from your time (noon), making it 7:00 A.M. in Honolulu. Once you get the hang of it, it's not hard at all.

THE INTERNATIONAL DATE LINE   The International Date Line, zigzagging through the Pacific Ocean from pole to pole just west of Alaska, looks as if it were drawn by an international committee, and it was. If you float over this line (most of it's over water) at, say, 9:00 A.M. on a Wednesday, going west, you leap twenty-four hours into the next day, to 9:00 A.M. on Thursday. Cross the line eastward and step into the day before. Confusing, yes?

INTERNATIONAL DATE LINE

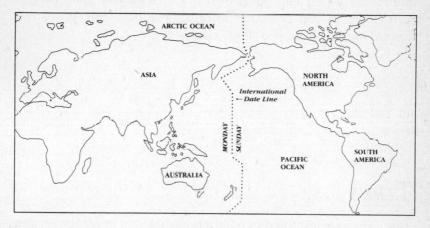

Source: Adapted from U.S. Bureau of Standards "Standard Time Zones of the World."

Without the International Date Line, however, it would be even more confusing. Look at a globe divided into twenty-four time zones (see **Time Zones** chart, page 185). It appears that all you need to do to get the local time in a far-off time zone is to count the time zones between you with your finger, hour by hour. Try it. Put yourself in Denver (Zone +7) at 7 A.M. on Thursday and try to find the local time in Tokyo (Zone −9). Going west, you'll count 8 time zones (+7 to ±12, then ±12 to −9), getting earlier all the time and arriving in Tokyo at 11 P.M. on Wednesday. Going east, however, you'll count 16 time zones (+7 to 0, then 0 to −9), getting later and later and finding it's 1 A.M. in Tokyo on Friday.

Confused? Try this: It's Thursday at 7 A.M., you're still in Denver (Mountain Standard Time), and you move west *all the way around the globe,* counting backwards by time zones as you go. If you do this, you'll find that it is 8 A.M. on *Wednesday* in Chicago, just one zone east of you. But if you move around the globe east from Denver, you'll find it's 6 A.M. on *Friday* in San Francisco!

To prevent this from happening, the International Date Line (the ± 12 zone) was established as the only place where there can be a day's difference between time zones. A day officially begins when it's noon in Greenwich, England, and midnight on the International Date Line. Because the earth spins eastward, when you cross the International Date Line going west toward Asia, you "gain" twenty-four hours. If you cross the International Date Line going east toward the Americas, you "lose" a day.

TIRES

There are so many different numbers stamped on tire sidewalls that one company manual requires fifteen pages just to introduce them to the sales personnel. Even if you actually figured out the sizing system, by the time you put 40,000 miles on a set of tires, the system had probably changed. In fact, tire sizing systems have changed *three times* since 1967. Unfortunately, remnants from the old systems have survived the new ones. With more than 2,000 types of tires on the market, you'll feel a lot better if you can figure out tire sizes and some of the quality and safety codes.

The "XH" is Michelin's code for a highway tire (each tire company has its own code). The other numbers are examined below. Keep in mind that these explanations are only guidelines—a sort of "Conversational Sidewall"—since little seems to be hard and fast in the tire industry. Tire manufacture is complicated—tires are not stamped out like Lifesavers. Familiarity with tire language, however, can at least give you a way to compare the possibilities.

CAN YOU DEAL WITH THIS TIRE AD?

XH-MICHELIN STEEL RADIAL WHITEWALLS		Michelin or Dunlop STEEL RADIALS		IMPORT STEEL RADIAL			
P185/80-13 **59.95**	P205/75-15 **68.95**	155-12	34.95	155-12	**19.95**	175-70-13	**28.95**
P185/75R-14 **59.95**	P215/75R-15 **71.95**	145-13	35.95	155-13	**22.95**	185-70-13	**29.95**
		155-13	37.95	145-13	**21.95**	185-70-14	**29.95**
P195/75R-14 **62.95**	P225/75R-15 **74.95**	165-13	40.95				
		175-14	48.95	175-13	**29.95**	195-70-14	**34.95**
205/75R-14 **66.95**	P235/75R-15 **76.95**	185-14	51.95	165-13	**24.95**	205-70-14	**37.95**
		165-15	46.95				
P215/75R-14 **69.95**		17570-13	50.95	185-14	**32.95**	165-15	**29.95**
		18570-13	53.95				
		18570-14	55.95	175-14	**29.95**	185-70-15	**59.95**
		19570-14	64.95				

Source: Courtesy of Consumer Tire Warehouse, Santa Rosa, Calif.

PASSENGER CAR TIRE SIZES

Example: P195/80R15 or just 195/80R15. Tire sizing systems have changed several times in the last thirty years, but since 1976 the most widely used system is the P-Metric System, created to accommodate the tiny tires used by the then-new small economy cars. The numbers the P-Metric System uses, however up-to-date the system is supposed to be, are an unbelievable hodgepodge: a metric number, a percentage representing a ratio, inches, and sometimes even a load limit number that isn't the load limit at all but a number *representing* the load limit. Here is a typical tire size: **P195/80R15.** It falls into 5 parts:

P means the tire is intended for a passenger vehicle. LT would mean "light truck," which may use a different size system (see below). Sometimes the P is left off.

195 is the section width, sidewall to sidewall, of the properly inflated tire *in millimeters.*

80 is the *aspect ratio* of the tire, or the relationship between the tire section height and the section width. It's a *percentage* figure. If a tire has an aspect ratio of 80, the height is 80 percent of the width. Essentially, a low-aspect ratio means the tire rides low and wide, which gives better control than a high ratio; e.g., 85, which rides higher but with more give.

R is the type of tire, in this case, radial. A B here means "belted bias," and a D signifies "diagonal bias." Diagonal bias tires are constructed with diagonally built layers; belted tires have additional layers that run around the tire, like a belt; radial tire layers run at right angles to a similar belt layer.

15 is the rim diameter *in inches* of the wheel the tire fits. The most common rim sizes are 13, 14, and 15 (whole numbers), and these wheel sizes are standard throughout the auto industry—it must really gall the very metric Europeans to have to size wheels in inches!

TIRE TYPES

BIAS   BELTED BIAS   RADIAL

Source: Tire Industry Safety Council.

Of the above numbers, you can select all but the rim diameter—if your car takes a 13-inch tire, you must buy a 13-inch tire. But you can change the width of the tire, or the type or aspect ratio. If the P-Metric system isn't used for the tires on your car, remember that a decimal is always decimal inches and almost always means tire width (sidewall to sidewall). If a size is preceded by a letter between A and N, it probably means load/size—the lower the letter, the smaller the size and load-carrying capacity of the tire.

LIGHT TRUCK TIRE SIZES
Example: 7.50R15LT or 31 × 10.50R15LT or LT215/75R15.
The sizes of light truck tires may be expressed in at least three (and probably more) ways, but LT always means "light truck."

1. **LT215/75R15:** This number reads the same as P tires.
2. **7.50R15LT:** The 7.50 here is the section width in inches instead of millimeters. The rest is the same as P tires.
3. **31 × 10.50R15LT:** The 31 is the diameter of the tire (not the rim) in inches, and the 10.50 is the section width (sidewall to sidewall) in decimal inches. The rest is the same as P tires.

LOAD INDEX

Example: *P195/60R15 86H*. Often found following the tire size number is another number-letter combination. The letter is the speed rating (explained in the next paragraph) and the number isn't 86 *of* anything but represents the maximum number of pounds the tire can safely carry when properly inflated. The load index range is 75 to 100, with the 75 representing 851 pounds and 100 representing 1,760 pounds.

SPEED RATING

Example: *195/60HR15 or 86H*. Speed ratings originated in Europe, where historically speed limits often have been higher than ours, if not absent. Europeans give tires one of four speed ratings, each represented by a letter. A word of warning: Although a high-speed rating may tell you something about a tire, all tires with the same rating are not equal. There are too many factors involved in building a tire to rely on one rating to judge it. There are two speed rating systems. In the first, the speed rating letter is usually found in the tire size number, inserted before the letter representing the tire construction, e.g., 195/60HR15.

Speed Rating	Top Speed of Vehicle
S	Up to 112 mph
H	Up to 130 mph
V	Up to 149 mph
Z	Over 149 mph

The second, more recent speed rating system uses only two letters and is found following the load index, e.g., **86H**.

P	Up to 93 mph
H	Up to 130 mph

SERIAL NUMBER Known as the DOT (Department of Transportation) number, the serial number is usually found on the opposite side of the tire from the size number, near the rim of the wheel.

(Tires are usually mounted with the serial number facing inside, so you may not be able to see it on your car's tires.) It is part of a system of tire identification that became federal law in 1971. A DOT number looks something like this: **DOT MAL9 ABC032.**

DOT	means that the tire meets or exceeds Department of Transportation safety standards.
MA	is the code number assigned by DOT to the manufacturing plant.
L9	is a code number for the tire size.
ABC	is a group of up to four symbols, optional with the manufacturer, to identify the brand or other significant characteristics of the tire.
032	is the date of manufacture, in this case meaning the third week (03) of 1972 (2).

Federal law stipulates that tire dealers must record the buyer's name and address with serial number(s) of the tire(s) purchased.

UNIFORM TIRE QUALITY GRADE LABELING
Example: Traction 150 Treadwear B Heat Resistance C. It may be helpful to know that the government now requires that tires have mileage and safety ratings stamped right on the tires. If you can't find the ratings on a tire, ask the dealer—reputable tire dealers have this information for any of their products. The system is called Uniform Tire Quality Grade Labeling (UTQGL), and is regulated by the National Highway Traffic Safety Administration of the U.S. Department of Transportation. As official as this sounds, there are widespread complaints in the tire industry that the system doesn't work, and that two tires with the same rating may show a wide difference in actual quality. According to some sources, one company's tires are not compared to another company's tires; instead, they are rated only against the tires offered by that company.

Also, a C rating on a tire may not make it a poor choice. Sometimes it's a trade-off—you may want open tread for good traction, but open tread heats up more easily, so you may have to settle for a heat resistance grade of C. Despite all the warnings, the numbers do tell you something about the tire and give you a way to begin comparing tires. Here is how the rating system works, according to the brochure called "Uniform Tire Quality Grading"* from the U.S. Department of Transportation, with some help from *The Car Book* (Jack Gillis, Harper & Row, published annually). All tests are supposed to be done on a specified government course and meet any other government specified controls.

Treadwear: 90 to about 330. To get the expected mileage, multiply the grade times 200. For example, a tire graded 100 will last for about 20,000 miles. The higher the rating, the farther the tire is expected to roll.

Traction: A, B, or C. An A-rated tire has the best traction, C the worst, representing the tire's ability to stop on wet pavement.

Heat Resistance: A, B, or C. An A-rated tire runs cool, while a C-rated tire tends to heat up faster. Hot tires wear down faster than cool tires.

OTHER SIDEWALL NEWS   Here's some more typical tire reading:

1. The name of the tire company
2. The country where the tire was manufactured
3. M/S label: whether tire meets regulations for a mud and snow tire; e.g., M+S
4. Whether the tire is tubeless or tube-type (with tube)
5. Ply information for tread and sidewalls, including the tire ply composition and materials used; e.g., plies: sidewall 2 polyester, tread 2 polyester + 2 steel + 2 nylon
6. Maximum load that can be carried at maximum pressure; e.g., max load 1,250 lbs. at 36 P.S.I. (pounds per square inch) max pressure
7. P-Metric sized tires are rated standard load (35 p.s.i.) or extra load (41 p.s.i.)
8. Safety warning, which may be stamped on the tire or found in the owner's manual.

*This booklet giving the quality ratings for more than 2,000 tires is available free from the U.S. Department of Transportation, National Highway Traffic Safety Administration, 400 Seventh Street SW, Washington, D.C. 20590.

T*YPE*

It was yet another Frenchman (the French have dominated the measurement scene), Pierre Fournier, who, in 1735, invented the point system to size up printers' type, and another Frenchman, François Didot, who based Fournier's points on the French inch, with 72 points to the inch. In 1886 American type founders adopted the French system but based the point size on the American inch, which was, at the time, slightly shorter than the French inch. To this day, the European and American point sizes do not agree.

Points measure the tallness of type, with type sizes running from a practically microscopic 1 point (now possible with computers) to the inch-high 72-point headline. The height of the typeface is measured from the lowest descender (such as the bottom of a "y") to the highest ascender (such as the top of an "h"), with a little leeway at the top and the bottom. In other words, type size is not actually a measure of the type, but of the space required to accept the tallest and lowest letters.

8 or 9 point type is commonly used in newspapers.
10 point type is used for many books.
12 point type is the larger typewriter size.

18 point type works for a small headline.

The space between lines, called *leading* (pronounced "led-ing"), is also measured in points. The width of a line of type is measured in *picas,* 6 picas to the inch. The width of columns—blocks of type—are measured in picas, as is the space between them and the margins, sometimes called "gutters." The pica type used on many typewriters was called that because each character was 1 pica wide (⅙ inch) and 12 points (⅙ inch) tall, and there were six lines of type to the inch. (Six lines/inch is standard for most typewriters, even those using the smaller "elite" type.)

Indentations are measured in *ems.* An em is about the size of the letter "M," hence the name, but is technically equal to the type's height. In other words, the size of an em for a 12-point typeface would be a sideways 12 points, but since points are used to measure only up and down, never sideways, the em is used to represent the sideways measure of the height. Most indentations are 1 or 2 ems.

This book demonstrates this: The type is one column 25 picas wide, set in ITC Garamond Book that is 10½ points high with 2½ points of leading and 1 em indentations. These numbers are crucial to a book's look and design and are usually chosen by the book's designer.

Unfortunately, figuring a type size from the printed page is all but impossible. The problem is twofold: First, there's no standard for the amount of leeway (or shoulder) at the top and bottom of the type. A larger shoulder makes a smaller letter, and vice versa. The only way to measure type is to measure the points in a block of solid lines with no leading between them and divide by the number of lines. Since almost all print contains leading, the opportunity to do this is impractically rare. Second, the width of the type (the number of characters per pica) can vary immensely—chunky, round letters with small ascenders will look much larger than lanky letters with long ascenders. The space between characters and words also varies. Note the variation in the following 12-point typefaces:

This is 12 point Helvetica type.

This is 12 point Times Roman type.

This is 12 point Clarinda Typewriter type.

This is 12 point Bodoni Book type.

This is 12 point Futura Book type.

This is 12 point Goudy Old Style type.

The only accurate way to gauge type for a printing job is to choose it from typeface sample books your printer will be happy to show you.

Universe (Distances)

The Astronomical Unit, the light year, and the parsec are the measurements used to estimate the distance between various points in the universe. These and other monstrous measurements offered by today's astronomers sound so terrifyingly huge that the mind resists them, convinced they are incomprehensible. The measures aren't hard to understand, though. They offer us a concept of how vast the universe is, as well as a healthy appreciation of size. The *Astronomical Unit (A.U.)* is the smallest—the distance between the earth and the sun, which can be measured in miles (93 million) or kilometers, or whatever you're comfortable with. The Astronomical Unit is useful for measuring distances in our solar system. A *light year* takes you into the galaxy. It's the distance light travels in one year—5,880,000,000,000 (or nearly 6 trillion) miles—at about 186,000 miles per second. The *parsec* is simply 3.2 light years.

It's interesting to note that light years represent the age of the image as well as its distance—the light from an object a light year away has taken a year to reach you, so you are seeing it as it was a year ago. The years you see into an object's past are numbered by its distance from you in light years: When you look at Rigel, a star in the constellation Orion, for example, which is 900 light years away,

Thumb-Sketch of the Universe

Description	Distance from the Sun	Distance on Your Chart Needed to Accommodate This scale: (1 A.U. = 1″ and 1 ly = 1 mi.)
The Earth	A.U. 1 Miles: 93 million	1 thumb-width OR 1 inch
Pluto, farthest planet from the sun	A.U. 29 Miles: 2.7 billion	29 inches
Distance light travels in one year	A.U. 63,310 Light years: 1 Miles: 6 trillion (5.88)	1 mile

Description	Distance from the Sun	Distance on Your Chart Needed to Accommodate This scale: (1 A.U. = 1″ and 1 ly = 1 mi.)
1 parsec: the distance light travels in 3.2 years	A.U. 200,000 Light years: 3.2 Miles: 19 trillion	3.2 miles
Alpha Centauri, one of our nearest stars	A.U. 270,000 Light years: 4.3 Miles: 25 trillion	4.3 miles
Rigel, a star in the constellation Orion	A.U. 57 million Light years: 900 Miles: 5.25 quadrillion	900 miles
M13 in Hercules, one of the farthest star groups visible to the naked eye in the N. Hemisphere	A.U. 1.7 billion Light years: 26,700 Miles: 160 quadrillion	26,700 miles*
The center of our galaxy	A.U. 2 billion Light years: 32,000 Miles: 190 quadrillion	32,000 miles
Nearest observed galaxies	A.U. 4.8 billion Light years: 75,000 Miles: 450 quadrillion	75,000 miles
Andromeda, our nearest full-size galaxy	A.U. 140 billion Light years: 2.2 million Miles 1.3 × 10^{19}**	2.2 million miles***
Nearest known quasar	A.U. 63 trillion Light years: 1 billion Miles: 6 × 10^{21}	1 billion miles
Distant quasars	A.U. 945 trillion Light years: 15 billion Miles: 9 × 10^{22}	15 billion miles

*Your chart must now be long enough to circle the earth.
See **Exponents, page 64.
***Your chart must now be 10 times longer than the distance from the Earth to the moon.

the image you see is 900 years old. Astronomers, staring deep into the Universe, are looking billions of years into the past.

How does the universe measure up? As it happens, there are almost the same number of inches in a mile (63,360) as there are Astronomical Units in a light year (63,310). An inch is traditionally about the width of your thumb. If you make your thumb represent 1 Astronomical Unit (the distance between the earth and the sun, or 93,000,000 miles), you can use it to chart the universe. The A.U. figures given in the chart on pages 196–197 represent how many times farther the distance is than that between the earth and the sun.

VISION

You know that 20/20 vision is excellent visual acuity. But what are those 20s twenty *of*? The answer is *feet.* You look at an eye chart from a distance of 20 feet. If you can read the lines normal eyes can see at 20 feet, you have 20/20 vision. If you can only read the lines people with normal vision can see at 40 feet, you have 20/40 vision. If your eyesight is better than normal, you might be able to read at 20 feet what most people have to get 5 feet closer to read, giving you 20/15 eyesight. The first 20 never changes.

This twenty-slash-whatever system is sometimes confusing. It is quite sensible to assume that the first number describes one eye, while the second describes the other; however, only the second number applies to you. It can describe your vision as you use both eyes, or each eye can be tested separately—one of your eyes might have 20/20 vision while the other has 20/40. So why don't they just drop the first 20?

Because there is another test for *up-close acuity* in which you are shown a card with lines of increasingly larger-sized characters from 14 inches away, the distance at which most people read. Normal up-close vision is described as 14/14. If you have 14/56 up-close vision, you need to get within 14 inches of reading material that a normally sighted person can see at 56 inches. The first number tells you whether the second number stands for your distance or your up-close acuity.

VITAMINS AND MINERALS

If you've ever tried to read the ingredients and figures listed on a bottle of vitamin tablets, especially those touted as high-potency, or even "super-high potency," you'd probably welcome a chart that explains it all. Unless your doctor is treating a special health problem, you may not need the impressive percentage over the U.S. RDA (United States Recommended Daily Allowance) some vitamin pills offer.

All vitamins are divided into two groups: those that are fat soluble—vitamins A, D, E, and K—and those that are water soluble—vitamins B and C. Your body stores the fat-soluble vitamins but gets rid of unused water-soluble ones, meaning that if you've been eating right, you've got a reserve of A, D, E, and K, but the C and B vitamins must be replenished every day. It's also far easier to get too much of the fat-soluble vitamins than the water-soluble ones.

It's easy to be put off by vitamin and mineral measures, which use both the metric system of grams (1 gram weighs about the same as a dollar bill), milligrams (one-thousandth of a gram), and micrograms (one-millionth of a gram) along with International Units. How much is an International Unit? There is no set value. An International Unit is the amount of a substance, such as a vitamin, that it takes to produce a certain effect when tested according to an internationally accepted biological procedure. In other words, if it takes more of one vitamin than of another to be effective, the value of one's International Unit will be higher than the other's. For example, it might take only 300 micrograms of vitamin A to produce a specified effect, but 1 whole milligram of vitamin E. The IU for that form of vitamin A equals 300 micrograms, and the IU for vitamin E equals 1 milligram.

To read a vitamin pill bottle, simply find the column of the following chart that applies to you (or your child) and compare it to that on the bottle. This information is more accurate than the U.S. RDA percentage figure found on many vitamin pill labels, which may not take the different requirements of children, men, and women into account.

RECOMMENDED DAILY ALLOWANCES (RDAS) FOR SOME VITAMINS AND MINERALS

Vitamin or Mineral	Infants and Children Through Age 10	Women	Men
A	1,400–3,000 IU	4,000 IU	5,000 IU
B1 (thiamine)	0.3–1.2 mg	1.0–1.1 mg	1.4–1.7 mg
B2 (riboflavin)	0.4–1.6 mg	1.2–1.3 mg	1.4–1.7 mg
B3 (niacin)	6–16 mg	13–15 mg	16–19 mg
B5 (pantothenic acid)	2–5 mg*	4–7 mg	4–7 mg
B6 (pyridoxine)	0.3–1.6 mg	1.8–2 mg	1.8–2.2 mg
B12 (cobalamin)	0.5–3 mcg	3 mcg	3 mcg
Biotin	35–85 mcg	100–200 mcg	100–200 mcg
Folic acid	30–300 mcg	400 mcg	400 mcg
C (ascorbic acid)	35–45 mg	50–60 mg	50–60 mg
D	400 IU	200–300 IU	200–300 IU
E	2–12 mg	10–20 mg	10–20 mg
Calcium	360–1,200 mg	800–1,200+ mg	800–1,200 mg
Phosphorus	240–800 mg	800–1,200 mg	800–1,200 mg
Magnesium	50–250 mg	300 mg	350–400 mg
Potassium	300–3,000 mg*	1,875–3,000 mg	1,875–5,625 mg
Iron	10–15 mg	10–18 mg (30–60 mg if pregnant)	10–18 mg
Zinc	3–10 mg	15 mg	15 mg
Selenium	0.01–0.12 mg*	0.05–0.2 mg	0.05–0.2 mg
Copper	1.0–2.5 mg	2.0–3.0 mg	2.0–3.0 mg
Manganese	0.1–2.5 mg*	2.5–5.0 mg	2.5–5.0 mg
Molybdenum	0.01–0.3 mg	0.15–0.5 mg	0.15–0.5 mg

*No RDAs have been established.

Based on information from Recommended Dietary Allowances. *Used with permission from the National Academy Press, Washington, D.C.*

There are too many minerals and trace elements to mention here, such as iodine, fluorine, chromium, and others, but the above are the most common.

Volume

The American method of measuring volume is one of the most antiquated measuring systems around, derived from the Egyptian mouthful and the twelfth-century "Winchester bushel," the oldest official English measure of volume on record. After that time, different measures were used for different goods for more than 500 years, and measures with the same names changed sizes depending on what was being measured. Thus a gallon of wine measured 231 cubic inches while a gallon of ale would measure 282 cubic inches, and so on. The most commonly used, however, were the Winchester bushel and the wine barrel.

These measures, along with the accompanying confusion, were transported to Colonial America. It is hard to believe that the mouthful, as well as the ancient custom of doubling each measure to find the next, had found their way to the New World:

2 mouthfuls	=	1 jigger
2 jiggers	=	1 jack
2 jacks	=	1 jill
2 jills	=	1 cup
2 cups	=	1 pint
2 pints	=	1 quart
2 quarts	=	1 pottle
2 pottles	=	1 gallon
2 gallons	=	1 peck
2 pecks	=	1 pail
2 pails	=	1 bushel
2 bushels	=	1 strike
2 strikes	=	1 coomb
2 coombs	=	1 cask
2 casks	=	1 barrel
2 barrels	=	1 hogshead
2 hogsheads	=	1 pipe
2 pipes	=	1 tun

It was obvious on both sides of the Atlantic that some sort of standard was necessary. In 1824 Britain decided to change the wine gallon to the imperial gallon, with a capacity of 277.5 cubic inches, and the bushel to 8 imperial gallons, or 2,219.4 cubic inches. The United States, however, opted for the old measures. In 1832 the U.S. Treasury officially adopted the old English wine gallon of 231 cubic inches and the even older Winchester bushel of 2,150.42 cubic inches. Even today, *none* of the British and United States measures of volume, though having the same names, have the same capacity—the British measures are slightly larger. This includes British cooking measures—their teaspoon, tablespoon, and cup are slightly larger than ours, nice to know if you're using a British cookbook. Britain has recently solved its problem of too many confusing volume measurements by adopting the metric system.

How do things stand in the United States? We have two official groups of volume measures, one for wet and one for dry. Some of these measures have the same names and some do not. The smallest wet measures, like the minum and the dram, are "apothecary" measures traditionally used for measuring drugs and medicines. The largest measure, the barrel, is rather imprecise—it can be as large as anyone says it is.

OFFICIAL U.S. MEASURES OF VOLUME, WET *

60 minums =	1 fluid dram
8 fluid drams =	1 fluid ounce
4 fluid ounces =	1 gill
4 gills =	1 pint
2 pints =	1 quart
4 quarts =	1 gallon

OFFICIAL U.S. MEASURES OF VOLUME, DRY *

2 pints =	1 quart
8 quarts =	1 peck
4 pecks =	1 bushel

*The barrel: The accepted volume of a barrel in the United States varies significantly, depending both on the commodity for which it is used and how it is defined in state law (varying from state to state). The barrel can measure both wet and dry volume.

Source: National Bureau of Standards, Department of Commerce.

Although the above measures are an improvement over those from Colonial times, the system is still flawed. Today's gallon and bushel are secondhand metric measures—both are defined in terms of cubic inches, which are by U.S. law defined in terms of the meter (see **The Metric System,** page 116). The simple metric solution, measuring volume by the liter, has still eluded us.

WEEK

Our seven-day week is purely arbitrary—it's not based on anything in nature—not the movement of the earth or anything in the heavens. The weeks are not really a division of anything, either—you can't group them evenly into months or even a year. The week somersaults awkwardly through the calendar with no quick way to tell which weekday will land on which date.

Technically, the week was probably inspired by a need to schedule market days and religious holidays. Its length has varied—the Egyptians and Greeks had a ten-day week, the Romans an eight-day week, China a fifteen-day week, and even today weeks from four to ten days are used by some of the world's more remote communities. Our seven-day week comes from the Jewish calendar, which bases its week on the seven-day creation, in which God worked for six days and rested on the seventh. Some have tried to outlaw the week: The French tried during the French Revolution, the Soviets tried it in the 1930s, but the week always bounced back.

Americans seem to love the week. It's user-friendly and scientists have refrained from trying to split the week into nanoweeks. We even increased our rest days from one day to two—the first and last day of the week—a precious, fairly recent period of time which we cherish as the weekend.

WEIGHT

Not just one but *three* perfectly legal U.S. weight systems involve pounds and ounces! What's the difference between a troy, an apothecary, and an avoirdupois pound? The answer is very old and simpler than you think.

The Grain. The seeds of our systems of weights were exactly that—seeds used with simple balances to weigh precious gems and metals and other small items. The ancient Egyptians used the grain as an official measure, in multiples of their favorite number:

ANCIENT EGYPTIAN WEIGHTS

60 grains	=	1 shekel
60 shekels	=	1 great mina
60 great minas	=	1 talent

The grain has prevailed through history, with values varying from half to a quarter of the value of today's U.S. grain, which is officially 0.0648 gram. This grain is the same for all U.S. ounce-pound systems—7,000 grains equal the familiar pound, and 5,760 equal the less familiar apothecary pound. In the beginning, though, the pound was smaller.

The Troy Pound. Ancient Egyptians have been credited with the first pound, which is said to have been $\frac{1}{100}$ of a cubic foot of water, or about 25 percent smaller than the pound you use today—5,244 grains. The Romans adopted the Egyptian pound, called it *libra pondo* (meaning "a pound by weight," hence our abbreviation *lb.*), and divided it into twelve parts of 437 grains each, calling each part an *uncia* (which means "twelfth part").

The use of this 12-ounce pound spread to Europe, where the ounce was enlarged to 480 grains and the pound was called the *troy pound*, honoring the thriving French medieval trading town of Troyes. When the troy pound reached England, it became popular for weighing precious metals and gems, and was adopted as the official weight system for British currency. The colonists brought the troy pound to America, where it was later officially adopted by the U.S Mint. The troy pound continued as the accepted measure of precious gems and metals on both sides of the Atlantic (hence its nickname "jeweler's weight"). Today the 12-

U.S. TROY WEIGHT

24 grains	=	1 pennyweight
20 pennyweights	=	1 ounce = 480 grains
12 ounces troy	=	1 pound troy = 5,760 grains

ounce troy pound has been all but abandoned for the metric system. It is still legal, however, and its equivalents are published by the U.S. Bureau of Standards.

The Apothecary Pound. In Europe the troy pound was also used by the medical world to measure drugs and medicines—the London College of Physicians adopted it in 1618—a custom carried on in America. Although an apothecary pound and ounce were and continue to be the same size as the troy pound and ounce, the apothecary scruple and the dram facilitated the mixing of small amounts. Although apothecary weights have been almost entirely replaced by the metric system, they are still legal in the United States and described by the Bureau of Standards:

U.S. APOTHECARY WEIGHTS

20 grains = 1 scruple
3 scruples = 1 dram = 60 grains
8 drams = 1 apothecary ounce – 480 grains
12 ounces = 1 apothecary pound = 5,760 grains

The Avoirdupois Pound. The everyday pound, the one you calculate your weight and your market produce in, is the avoirdupois pound. The avoirdupois system of weights is yet another rearrangement of the troy system. When the 5,760-grain troy pound was considered by many medieval Europeans to be too small for everyday use, they added four troy-size (480 grain) ounces to make it a 16-ounce pound of 7,000 grains. (The literal meaning of ounce is "12th-part" even if it has become a 16th-part.) The *avoir-*

AVOIRDUPOIS WEIGHTS
(pronounced "a-ver-de-**poiz**")

$27^{11}/_{32}$ grains = 1 dram
16 drams = 1 avoirdupois ounce = 480 grains
16 ounces = 1 pound = 7,000 grains
100 pounds = 1 hundredweight
20 hundredweights = 1 ton* = 2,000 pounds

*The 2,000-pound ton, called the *short ton,* is the common U.S. ton. For some purposes, however, we use the British ton of 2,240 pounds avoirdupois, called the *long ton.* The increasingly used metric ton equals 2,204.623 pounds.

dupois pound (French for *goods of weight*) became the accepted general measure of weight in both England and America.

The three systems have many similarities. Although the troy and apothecary pounds contain 12 ounces while the avoirdupois pound contains 16, and the apothecary and avoirdupois dram-size differs, all three systems use the 0.0648-gram grain and the 480-grain ounce. They have something else in common: The metric system has, directly or indirectly, replaced them all—the U.S. avoirdupois pound has been officially defined in terms of the kilogram (equaling 0.453 592 37 kilogram) since 1893, while most of today's experts in medicines and precious metals and stones have replaced their drams, scruples, and pennyweights with grams and milligrams. (See **The Metric System,** page 116.)

WIND

Most people have figured out that when the wind direction is reported on the news, it is always the direction the wind is coming *from.* For example, a SW wind is blowing from the southwest (the weather vane arrow always points *into* the wind). Wind direction is often a good weather predictor—even Aristotle observed that in his neck of the world, winds from the northwest brought clearing weather, those from the northeast brought snow and cold, while south and east-southeast winds were hot.

Wind speed, known as *velocity,* is measured by an *anemometer*— Robert Hooke invented the first one in 1667—which today most commonly consists of three cups spinning around a vertical shaft; the wind speed is measured by the speed of the spin. Lacking such sophisticated instruments, most of us rely on weather news, which reports the wind speed in *miles per hour,* a difficult figure to translate into reality—how forceful is a 30 mph wind anyway? The answer to this very sensible question was crucial to farmers and sailors, so as early as 1806, an admiral in the British Navy, Sir Francis Beaufort, devised a scale which, only slightly revised, is still in use today.

BEAUFORT'S WIND SCALE

Wind Speed in mph	Beaufort Number	Effects on Land	Official Designation
Less than 1	0	Calm; smoke rises vertically	Light
1–3	1	Smoke shows wind direction	
4–7	2	Wind felt on face, leaves rustle, wind vanes move	
8–12	3	Leaves, small twigs move; small flags extend	Gentle
13–18	4	Wind raises dust and loose paper; small branches move	Moderate
19–24	5	Small leafy trees sway; small waves crest	Fresh
25–31	6	Large branches move; umbrellas become unwieldy	Strong
32–38	7	Whole trees sway; walking into the wind is difficult	
39–46	8	Twigs break from trees; cars veer	Gale
47–54	9	Slight structural damage	
55–63	10	Trees are uprooted, a good deal of structural damage	Whole Gale
64–72	11	Widespread damage	
73 up	12	Widespread damage	Hurricane

You might assume that the force (or destructive capability) of a wind increases equally with its velocity (speed). But although the speed of a 40 mph wind is twice that of a 20 mph wind, its force is equal to that two times *squared* (2 × 2), meaning that a 40 mph wind is four times more powerful than a 20 mph wind. If a 30 mph wind is three times faster than a 10 mph wind, its force is that three times *squared* (3 × 3) making it nine times more powerful than the 10 mph wind. This makes a 40 mph wind *sixteen* times more powerful than a 10 mph wind (4 squared, or 4 × 4).

ZIP *CODES*

In 1963 the U.S. Postal Service numbered every American post office with a 5-digit ZIP Code. The numbers began with zeros at the farthest point east—00601 for Adjuntas, Puerto Rico—and worked up to nines at the farthest point west—99950 for Ketchikan, Alaska. It seemed outrageous at the time—the ZIP Code's *raison d'être* was facility of computer sorting, and computers were not yet familiar to most Americans. In 1983 the U.S. Postal Service added four digits to the ZIP Code, and as common as computers are these days, this seems outrageous. Most people still don't know their 9-digit code, and those who do resist using it.

It really is a fairly efficient system, however, and why fight anything that will speed up the mail? Here's what the first 5 digits mean. Let's use the ZIP Code of Healdsburg, California, as an example:

95448

9   The first digit represents one of 10 geographical areas, usually a group of states. The numbers begin at farthest points east (0) and end at the farthest points west (9).

54   The second two digits, in combination with the first, identify a central mail-distribution point known as a sectional center. The location of a sectional center is based on geography, transportation facilities, and population density; although just 4 centers serve the entire state of Utah, there are 6 of them to take care of New York City.

48   The last two digits indicate the town, or local post office. The order is often alphabetical for towns within a delivery area— for example, towns with names beginning with "A" usually have low numbers. (This may not apply to metropolitan areas, where numbers are assigned as they become available.)

The last four digits added after a dash, e.g., 95448-1234, are called ZIP + 4 coding. Mail with ZIP + 4 coding benefits from cheaper bulk rates, being easier to sort with automated equipment. It's also helpful for businesses which wish to sort the recipients of their mailings by geographical location. The first two numbers of the 4-digit suffix represent a delivery sector, which may be several

ZIP CODE NATIONAL AREAS

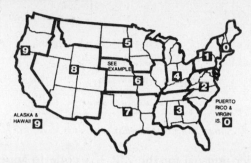

The first digit of a ZIP Code divides the country into 10 large groups of states numbered from 0 in the Northeast to 9 in the far west.

EXAMPLE

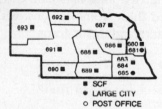

- ■ SCF
- ● LARGE CITY
- ○ POST OFFICE

Within these areas, each state is divided into an average of 10 smaller geographic areas, identified by the 2nd and 3rd digits of the ZIP Code.

WHAT YOUR ZIP CODE MEANS

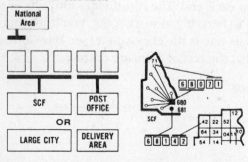

The 4th and 5th digits identify a local delivery area.

Source: U.S. Postal Service 1988 National Five-Digit ZIP Code and Post Office Directory. Used with permission.

blocks, a group of streets, several office buildings, or a small geographical area. The last two numbers narrow the area further: They might denote one floor of a large office building, a department in a large firm, or a group of post office boxes.

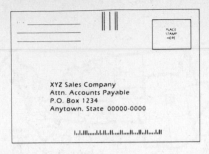

Source: U.S. Postal Service.

It is no longer unusual for the ZIP + 4 code to appear as a bar code at the bottom right corner of the envelope. This is for computers, of course (see **Computers**, page 43). It's easy to read this, actually. There are 10 combinations of 5 bars (the vertical lines in the above illustration), each containing two long bars, and three short. Digits 0 to 9 have been assigned these combinations, with long bars standing for ones and short bars standing for zeros.

0	11000	3	00110	7	10001
1	00011	4	01001	8	10010
2	00101	5	01010	9	10100
		6	01100		

The following bar code represents the ZIP Code 12345-6789. There is a tall frame bar at each end. The rest divides into sets of 5-bar short (0's) and tall (1's) bars. If you work it out, you'll see there are 10 digits, not 9. The last one is the check digit (see **Bar Codes**, page 7), which ensures that the rest of them are correct.

ZIP + 4 BAR CODE

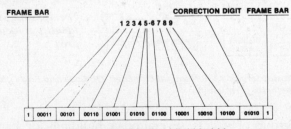

Source: U.S. Postal Service.

BIBLIOGRAPHY

Adams, Catherine F. *Nutritive Value of American Food in Common Units.* Washington, D.C.: U.S. Department of Agriculture, 1975.

Agricultural Marketing Service. Washington, D.C.: U.S. Department of Agriculture, 1986.

American Institute of Real Estate Appraisers. *The Appraisal of Real Estate,* 8th edition. Chicago: American Institute of Real Estate Appraisers, 1983.

Asimov, Isaac. *Asimov on Numbers.* New York: Doubleday & Co., 1977.

———. *The Clock We Live On.* New York: Abelard, Schuman, 1965.

———. *Realm of Numbers.* Boston: Houghton Mifflin Co., 1959.

———. *The Universe,* revised edition. New York: Walker & Co., 1980.

Berger, Melvin. *For Good Measure.* New York: McGraw-Hill, 1971.

Bishop, Owen. *Yardsticks of the Universe.* New York: Peter Bedrick Books, 1984.

Boorstin, Daniel J. *The Discoverers.* New York: Random House, 1985.

Branley, Franklyn M. *Think Metric!* New York: Crowell, 1973.

Brondfield, Jerome. "The Marvelous Marking Stick." *Kiwanis Magazine,* February 1979.

Bureau of Labor Statistics Handbook of Methods, Volume II: The Consumer Price Index. Washington, D.C.: U.S. Department of Labor, April 1984.

Burns, Marilyn. *About Time.* Boston: Little, Brown, & Co., 1978.

Busha, Ed, et al. *The Book of Heat.* New York: The Stephen Greene Press, 1982.

Campbell, Tim. *Do-It-Yourself Weather Book.* Birmingham, Ala.: Oxmoor House, 1979.

Consumer Tire Guide. Washington, D.C.: Tire Industry Safety Council, n.d.

Crocket, James Underwood. *Vegetables and Fruits.* Alexandria, Va.: Time-Life Books, 1972.

Diagram Group. *Comparisons.* New York: St. Martin's Press, 1980.

Encyclopedia Americana, international edition, Danbury, Conn.: Grolier, 1987.

Fundamental Facts About United States Money. Atlanta: Federal Reserve Bank of Atlanta, 1986.

Food and Nutrition Board. *Recommended Dietary Allowances,* 9th revised edition. Washington, D.C.: National Academy of Sciences, 1980.

Forest Products Laboratory. *Wood Handbook: Wood as an Engineering Material,* revised edition. Washington, D.C.: U.S. Department of Agriculture, 1974.

Gallant, Roy A. *Man the Measurer.* New York: Doubleday & Co., 1972.

Gillis, Jack. *The Car Book,* 1987 edition. New York: Harper & Row, 1987.

Gioello, Debbie Ann. *Figure Types and Size Ranges.* New York: Fairchild Books, 1979.

Goudsmit, Samuel Abraham, and Robert Claiborne. *Time.* Alexandria, Va.: Time-Life Books, 1980.

HELP! The Indispensable Almanac, 1986 edition. New York: Everest House, 1986.

The Heritage of Mechanical Fasteners. Cleveland, Ohio: Industrial Fasteners Institute, 1974.

Hirsch, S. Carl. *Meter Means Measure.* New York: The Viking Press, 1973.

Historical Statistics of the United States, Colonial Times to 1970. Washington, D.C.: U.S. Department of Commerce, 1975.

Housman, Patricia. *The Calcium Bible.* New York: Rawson Associates, 1985.

Jespersen, James, and Jane Fitz-Randolf. *From Sundials to Atomic Clocks.* Washington, D.C.: National Bureau of Standards, U.S. Department of Commerce, 1977.

Judson, Lewis V. *Weights and Measures Standards of the United States,* revised edition. Washington, D.C.: U.S. Department of Commerce, 1976.

Lewis, Edward V. *Ships,* revised edition. Alexandria, Va.: Time-Life Books, 1973.

McDonald's Food: The Facts. Oak Brook, Ill.: McDonald's Nutrition Information Center, 1987.

A Manual of Style, 13th edition. Chicago: University of Chicago Press, 1985.

Mayes, Kathleen. *Osteoporosis.* Santa Barbara, Calif.: Pennant Books, 1986.

McGraw-Hill Encyclopedia of Science and Technology, 6th edition. New York: McGraw-Hill, 1987.

Miller, Albert, et al. *Elements of Meteorology.* Columbus, Ohio: Charles E. Merrill, 1983.

Miller, Peter. *The Common Sense Mortgage.* New York: Harper & Row, 1985.

Moroney, Rita L. *The History of the U.S. Postal Service 1775–1984.* Washington, D.C.: U.S. Postal Service, n.d.

Motor Gasolines. Richmond, Calif.: Chevron Research Company, 1985.

National Oceanic and Atmospheric Administration. *Tide Tables 1987, West Coast of North and South America.* Washington, D.C.: U.S. Department of Commerce, 1987.

National Bureau of Standards. *What About Metric?,* revised edition. Washington, D.C.: U.S. Department of Commerce, 1985.

———. *Units and Systems of Weights and Measures: Their Origin, Development and Present Status.* Washington, D.C.: U.S. Department of Commerce, 1985.

National Weather Bureau. *The Aneroid Barometer.* Washington, D.C.: U.S. Department of Commerce.

The New Encyclopaedia Britannica. Chicago: Encyclopaedia Britannica, Inc. 1986.

Osol, David C., editor. *Remington's Pharmaceutical Sciences,* 16th edition. Easton, Pa.: Mack, 1980.

Oxford English Dictionary. New York: Oxford University Press, 1971.

Paul, Henry E. *Binoculars and All-Purpose Telescopes.* Radnor, Pa.: Chilton Book Co., 1965.

"Reading the Sidewall," *Modern Tire Dealer,* 1983.

Reithmaier, L. W. *Private Pilot's Guide.* Fall River, Mass.: Aero, 1977.

Rossi, William A. "How Shoe Sizes Grew," *Footwear News Magazine.* New York: Fairchild Publications, Summer 1988.

Saitzyk, Steven L. *Art Hardware.* New York: Watson-Guptill, 1987.

Staplers and Paper Fasteners. Alexandria, Va.: National Office Products Association, 1981.

Swezey, Kenneth M. *Formulas, Methods, Tips and Data for Home and Workshop.* Albany, N.Y.: Popular Science, 1969.

Thomas, Robert B. *The Old Farmer's Almanac.* Dublin, N.H.: Yankee Books, 1986.

Thompson, Philip D. *Weather.* New York: Time-Life Books, 1965.

Turbak, Gary. "Measuring Our Metric World (Minus America)." *Kiwanis Magazine,* October 1987.

U.S. Surgeon General. "The Changing Cigarette." Washington, D.C.: U.S. Department of Health and Human Services, 1981.

Uzes, Francois D. *Chaining the Land.* Rancho Cordova, Calif.: Landmark Enterprises, 1979.

Vivian, John. *Wood Heat.* Emmaus, Pa.: Rodale Press, 1976.

Walker, Bryce, and the editors of Time-Life Books. *Earthquake.* Alexandria, Va.: Time-Life Books, 1982.

Wilford, John Noble. *The Mapmakers.* Alfred A. Knopf, 1981.

YMCA of the U.S.A. *The Official YMCA Fitness Programs,* edited by the YMCA Board under the direction of William B. Zuti. New York: Rawson Associates, 1984.

World Book Encyclopedia. Chicago: World Book, Inc., 1987.

INDEX

Abrasive paper, 155–56
Account number, bank, 23–24
Acidity. *See* pH.
Acre, 106, 120–21
Addresses:
    street, 173–74
    ZIP Codes for, 208–10
Alcohol, 5–7
    blood levels of, 6–7
    in beverages, 5–6, 112
Altitude, effects of:
    on barometers, 14–15
    on engine octane requirements,
      87
Amp, 60, 116
Anemometer, 206
Angles, 30–31
Angstrom, 151
Annual Percentage Rate (APR), 7
Anthropometric studies, 33
Apothecary measures, 202, 205–6
Appliances:
    electric, 62–63
    gas, 84
    microwave, 123, 150
Area, formulas for:
    circles, 32–33
    cylinders, 126
    houses, 57
Aspect ratio for tires, 189–90
Astronomical Unit, 196–98
Atomic time, 181, 183–84
Atomic weight, 4
Atto- (prefix), 142–43
Attogram, xiv, 142–43
Automobile(s):
    Crash Test Rating Index, 51
    engines, 124–26
    gasoline, 85–89
    "knocking" problems, 85–86
    oil, 128–29
    tires, 188–93
Avoirdupois weight system, 205–6

Babylonians, 31
Bar codes, xiii–xiv, 7–11
    for ZIP+4 codes, 210
Barleycorn, 161
Barometer, 12–15
    adjusting for altitude, 14–15
Barometric pressure, 12–15
    wind barometer table, 13–14
Barrel, 201–2
    of oil, 88

Beaufort, Sir Francis, 206–7
  wind scale, 207
Beer, xv, 5–6
  light, 112
  pH of, 134
Bias tires, 189–90
  belted, 189–90
Binary number system, xiv, 9–10,
  44
Binoculars, 15–16, 116
Bit (binary digit), 44–45
Blood pressure, 16–17, 116
Bolts, 156–58
Book:
  copyright page for, 49–51
  ISBN number for, xiv, 50, 105
  publishing history of, 50
  typeface for, 194–95
British measures, 55, 162, 201–2
Btu, 83–85
Bushel, 120–21, 201–3
  Winchester, 201
Byte, 44–45

Cable, 55
Calcium, 17–19
  in foods, 18
  RDAs for, 200
  supplements, 19
Calendar, 19–22
  World Calendar Association, 21–
  22
Calorie (measure of heat), 78–79,
  85
Calories (kilo-), 78–81, 85, 116
  in beer, 112
  expended in exercise, 80
Cancer, skin, 175
Cans, 22
Carat, 88, 116, 140
Carbon-14 dating, 4

Carbon monoxide in cigarettes,
  26–30
Carburetor, 86, 125
Celcius, Anders, 175–76
  temperature scale, 175–77
Centi- (prefix), 119
Centigrade temperature scale,
  175–77
Centimeter, 116–21
  cubic, 119, 124–26
  square, 122
Chain, 145
Chaldeans, 54
Check digit:
  in bank account numbers, 24
  in bar codes, 8, 10–11
  in ISBN numbers, 105
  in ZIP+4 bar codes, 210
Checks, bank, 23–24
Cholesterol, 17–18, 24–26, 116
  blood levels of, 24–25
  effect of exercise on, 81
  in foods, 25–26
Cigarettes, 26–30
  light, 26–30
Circles, 30–32
  circumference of, 32
City planning, 173–74
Clocks, 30, 171
  atomic, 183–84
Clothing sizes, 33–37
  babies', 33–34
  children's, 34–35
  hats, 92–93
  men's, 35–36
  shoes, 161
  socks, 164–65
  women's, 36–37
  XS–XL, 33, 36
Compass, 41–43
Computer:
  effects of, xiv

memory, 43–46, 141
See also Binary number system.
Consumer Price Index, 46–49, 73
Conté, Nicolas Jacques, 131
Coordinates, map, 108–110
Copyright page, 49–51
Cord of wood, 74–76
    face, 75
Cost of living index. *See* Consumer
    Price Index.
Crash Test Rating Index, 51
Cup, 121, 201
Currency (notes), 51–54
Cylinder:
    area of, 126
    in engines, 125–26

Daimler, Gottlieb, 125
Dating prehistoric objects, 3–4
Day, 20, 181, 183–84
    lunar, 177
Daylight saving time, 182
    adjusting tide tables for, 179
Deci- (prefix), 119
Decibel, 170–72
Decimal foot, 146
Decimal inch, 12, 111
Declination, U.S., 42–43
Degree:
    circle, 31
    compass, 41
    latitude and longitude, 31, 107–
        110
    temperature, 37–40, 175–77
Deka- (prefix), 119
Diastolic blood pressure, 16
Didot, François, 194
Diet, 78–81
    balanced, 79
    vitamins and minerals in, 199–
        200
    See also Calcium; Cholesterol;

Exercise; Food, Sodium; and
    Vitamins and minerals.
Distances:
    astronomical, 196–98
    nautical, 54–55
    See also Length; names of indi-
        vidual units of measure.
Dow, Charles, 56
Dow Jones Averages, 55–56
Dozen, 40–41
Dram, 202, 205–6

Earth, age of, 4
Earthquakes, 58–60
Ebb current, 180
Edward I, King of Great Britain,
    110
Edward II, King of Great Britain,
    161
Egypt, 31, 110, 136, 203–4
Electricity, 60–63, 116
    light bulbs, 112–14
    wavelengths of, 150
Elements (chemical), 3–4
Elizabeth I, Queen of Great Britain,
    54, 110
Ell, 67
Em, 195
Engines:
    action of (four-stroke), 125–26
    horsepower of, 63–64, 124
    octane requirements of, 86–88
    size of, 116, 124–26
    steam, 63
Ephemeris second, 183
Equator, xiv, 107–9
Erastosthenes, 107
Exercise, 80–81
    as affects calcium, 19
    calories expended during, 80
    optimum heart rate for, 94–95
Exponents, 64–66

Fabric:
  care labels, 66–67
  widths, 67–69
Fahrenheit, Gabriel, 175
  temperature scale, 175–77
Fathom, 55
Federal Reserve Board, 51–53, 73, 144
  districts, 52–53
Femto- (prefix), 142
Fertilizers, 69–72, 167
Fiber, dietary, 17
Financial indexes, 73–74
Firewood, 74–78
Flood current, 180
Food:
  calcium in, 18
  calories in, 78–81, 116
  cholesterol in, 25
  grading of, 81–82
  sodium in, 165–66
  See also Diet; Produce, fresh.
Foot (unit of measure):
  board, 115
  cubic, 121, 123, 159–60
  decimal, 146, 179
  history of, xiii–xiv, 110
  linear, 114
  metric equivalent for, 118, 120–122
  square, 57, 106, 122
Foot-pound, 63
Fossils, 3–4
Fournier, Pierre, 194
Fractions, 111
Franklin, Benjamin, 139–40
Frequencies:
  electromagnetic waves, 149–51
  radio, 116
  sound, 170–72
Furlong, 106, 145

Galileo, 12
Gallon, 201–3
  metric equivalent for, 117, 120–122
Gamma rays, 150–51
Gardening. *See* Fertilizers; Soil (garden).
Gas, natural, 82–85
Gasoline, 85–89
Gauges:
  nail, 127
  rain, 152
  screw and bolt, 127, 157
  snow, 162
  wire, 127
Giga- (prefix), 141
Gill, 202
Gold, 88–89
  element, 3
Gowd, Charles, 184
Grain, 120–21, 204–6
Gram, 119–22, 199, 206
Greenwich Mean Time (GMT), 89, 108, 123–24, 182, 184
Gregory XIII, Pope, 21
Gross, 41
  great, 41
Gross National Product, 73, 89–92
Gunther's chain, 145

Half-life, 3–4
Hats, 92–93
Hearing. *See* Sound.
Heart rate, 93–95
Heat:
  measure of. *See* Btu; Calorie.
  resistance, 99–104
  for home, 84
Heat stress index, 39–40
Hectare, 120–21
Hecto- (prefix), 119

Hertz, 116, 150–51, 170
Hide, 106
Highway signs, 95–99
    distance, 96
    interstate routes, 96–97
    mileposts, 98
    state and county routes, 98
    U.S. routes, 97–98
Hipparchus, 107
Hooke, Robert, 206
Horsepower, 63–64, 124
Houses:
    area of, 57
    insulating, 99–104
Humidity, relative, 39–40, 99
Humiture, 39–40
Hundredweight, 205
Hydrocarbons, 87
Hygrometer, 99

Inch, xiii, 111
    cubic, 120, 202–3
    decimal, 12, 111
    metric equivalent, 120–22
    micro-, 111, 142
    square, 122
Indexes:
    Consumer Price Index, 46–49,
        73
    financial, 73–74
    Gross National Product, 73, 89–
        92
Insulation for homes, 99–104
    types of, 104
Interest:
    on loans, 137–38
    Prime Rate, 143–44
International Date Line, 184–88
International Unit, 199–200
ISBN, xiv, 50, 105

Isomer, 87
Isotope, 4

Jewelry, gold, 88–89
Jewels, 117, 140
Jones, Edward, 56

K (kilobyte), 44–46
Karat, 88–89
Kelvin, Lord William Thomson,
    176–77
Kelvins, 176–77
Kilo- (prefix), 119, 141
Kilobyte, 44–46
Kilocalorie. *See* Calories (kilo-).
Kilogram, 117–22, 206
Kilohertz, 150–51
Kilometer, 120–22
    square, 122
Kilometers per hour, 121–22
Kilowatt hour, 61–62, 141
Knot (nautical measure), 54–55
Krypton-86, 111

Land, 106
    surveying, 145–49
Latitude and longitude, 31, 107–
    110
    affecting nautical measures, 54–
        55
    affecting time zones, 184–86
Lead:
    element, 3–4
    printer's, 194–95
Leading Indicators index, 73
Leap-second, 181, 184
Leap-year, 21
Length, xiii–xiv, 110–11, 117
    metric measures, 116–22
    See also names of individual
        units of measure.

Library of Congress Cataloging in Publication (CIP), 50
Light:
  and meter definition, xiv, 111, 117
  brightness of, 113
  speed of, 150
  wavelength of, 151
Light bulbs, 60, 112–14
Light year, 64, 196–98
Link, 145
Lipoproteins, 25
Liquor, 5–6
Liter, 116–22, 202–3
Loans:
  interest on, 143–44
  points on, 7, 137–38
Logarithmic scales:
  decibel, 170–72
  exponents, 64–66
  pH, 133
  Richter, 59
Longitude. *See* Latitude and Longitude; Meridian(s).
Lot and block surveying, 149
Lumber, 114–16
  grades of, 115–16
  plywood, 136–37
Lumen, 113

Mass. *See* Weight.
Mach, 172
Mean solar time, 181
Medicines, 116. *See also* Apothecary measures; International Unit.
Mega- (prefix), 141
Megabyte, 44–45
Megahertz, 150–51
Mercalli, Giuseppe, 58
  earthquake scale, 58

Mercury:
  in barometers, 12
  in sphygmomanometers, 16–17
Meridian(s), 107–9
  Prime, 107–9
  principal, 147
Mesopotamians, 31
Meter, xiv, 111, 117–22
  cubic, 121
  square, 122
Meters (instruments):
  gas, 82–83
  electric, 61
Metes and bounds surveying, 146–47
Metric system measures, xiii–xiv, 116–22
  for electromagnetic waves, 149–151
  history of, xiii–xiv
  for length, 116–22
  P-Metric tire sizing, 189–90, 193
  prefixes for, 118–19, 141–43
  for shoe sizes, 161–62
  for temperature, 175–77
  in U.S., xiv, 116–18, 206
  for vitamins and miernals, 199–200
  for volume, 202–3
  for weight, 117–22
Micro- (prefix), 142
Microgram, 199–200
Microwave ovens, 123, 150
Mil, 111
Mile:
  history of, xiv
  nautical, 54–55, 122
  square, 106
  statute, 54, 120–22, 145
Miles per hour, 121–22
  wind speed, 206–7
Military time, 89, 123–24
Milli- (prefix), 119, 142

Millibar, 12
Milligram, 116, 119–21, 199–200
Milliliter, 120, 122
Millimeter, 116, 119–21
Minum, 202
Minute:
    in a degree, 31, 109–10
    of time, 31, 183
Money market, 144
Month, 20–21
Mortgages. *See* Loans.
Motorcycles, 124–26, 177

Nails (hardware), 126–28
Nano- (prefix), 142
Nanosecond, 182–84
National Earthquake Information
    Center, 59
Nautical measures, 54–55, 122
Neanderthal man, 4
New York Stock Exchange, 55–56,
    131
Nicotine in cigarettes, 26–30, 116
Nitrogen in fertilizers, 69–72
North:
    magnetic, 41–43
    pole, xiv, 107–8, 111, 117
    true, locating, 43
Number systems
    binary, 9–10, 44, 111
    Roman, 152–54
    sixty-based, 31
    ten-based, 40. *See also* Metric
        System measures.
    twelve-based, 40
Numbers:
    negative, 65–66
    proliferation of, xiii–xv
    very large, 64–66, 141
    very small, 64–66, 141–43

Octane, 85–89
Ohm, 60–61, 116

Oil:
    crude, 88
    engine, 128–29
Osteoporosis, 17. *See also* Cal-
    cium.
Otto, Nicholas, 125
Ounce, 203–6
    dry, 122
    fluid, 120, 122, 202
    metric equivalent for, 120–22

Pace, xiii, 54
Paper, 129–30
    abrasive, 155–56
Paper clips, 130–31
Parallels (latitude), 107–10
Pascal, Blaise, 15
Parsec, xiv, 196
Peck, 201–2
Pencils, 131–32
Penny (nail sizes), 126–27
Pennyweight, 204–6
pH, 132–35
    of garden soil, 166–69
Phon, 170
Phosphorus:
    in fertilizers, 69–72
    RDAs for, 200
Pi, 32–33, 126
Pica, 194–95
Pico- (prefix), 142
Pins (sewing), 135–36
Pint, 201–2
    dry, 121
    fluid, 120
    metric equivalent, 120–22
Pitch:
    of screws and bolts, 157
    of sound, 170
Plywood, 136–37
P-Metric tire sizes, 189–90,
    193

Points:
  on loans, 7, 137–38
  on stock market, 56
  of type, 194–95
Postal rates, 138–40
Potassium:
  -40 dating, 3
  in fertilizers, 69–72
  RDAs for, 200
Pound, 117, 203–6
  apothecary, 205–6
  avoirdupois, 205–6
  metric equivalent for, 117, 119–122
  paper, weight of, 129–30
  troy, 204–5
Precious stones, 116, 140
Prefixes, metric, 118–19, 141–43
Prehistoric objects, dating of, 3–4
Prime Meridian, 107–9
Prime rate, 143–44
Produce, fresh:
  grading of, 81–82
  pH of, 134
  sizing of, 144–45
Proof of alcohol, 6
Property, 106, 145–49
Pulse, 93–95

Quart, 201–2
  dry, 120
  fluid, 120
  metric equivalent for, 118–22

Radio, 116
  frequencies, 149–51
Radioactive elements, 3–4
Rain, 13–14, 151–52
RAM computer memory, 45–46
RDAs (U.S. Recommended Daily Allowances), 199–200
Ream of paper, 129–130

Rectangular surveying, 147–48
Richter, Charles F., 58–59
  earthquake scale, 59
Rod, 106, 145
ROM computer memory, 45–46
Roman numerals, 152–54
Rubber bands, 154–55
R-values, 99–104

Salt, 165–66
Sandpaper, 155–56
Scanners (UPC), 11
Screws and bolts, 156–58
Scruple, 205–6
Sea level, 180
Second:
  in degrees, 31, 109–10
  ephemeris, 183
  leap, 181, 184
  of time, 31, 142, 183–84
Section of land, 147–48
Seismograph, 59
Share of stock, 56
Ships, 159–61
Shoes, 161
Sidereal time, 181
Simpson, Edwin B., 161
Sixty, 31
Smoking. *See* Cigarettes.
Snow, 162
Social Security numbers, 162–64
Socks, 164–65
Sodium, 3, 165–66
Soil (garden):
  pH of, 166–69
  testing, 168–69
Solar time, apparent, 20–21, 181, 184
Sorenson, Soren, 133
Sound, 170–72
  speed of, 172

Speed:
of light, 64, 196–97
rating for tires, 191
of sound, 172
of winds, 207
SPF (Sun Protection Factor), 174–175
Sphygmomanometer, 16–17
Staples, 172–73
Steel wool, 173
Stock market, 55–56
Sundial time, 181, 184
Sunscreen lotion, 174–75
Supersonic speeds, 172
Surveyor's measures, 145–46
Systolic blood pressure, 16

Tablespoon, 122, 202
Tar in cigarettes, 26–60, 116
Teaspoon, 122, 202
Television channels, 149–51
Temperature, 175–77
Tera- (prefix), 141
Therm, 83–84
Tide(s), 177–80
mean low, 180
mean lower low, 180
neap, 180
spring, 180
tables, 177–80
Time, 108, 123–4, 181–88
International Date Line, 184–88
24-hour time, 123–24
types of, 181–82
units, 183–84
zones, 89, 184–88
See also names of individual units and types.
Tires, 188–93
quality of, 192–93
types of, 189–90

Ton:
dead weight, 160–61
displacement, 160–61
gross, 159–60
long, 121, 160, 205
metric, 121–22, 205
register, 159–60
short, 121–22, 160, 205
Torricelli, 12
Township, 147–48
Troy weight system, 88, 204–6
Tun, 159, 201
20/20 vision, 198
Type (printer's), 194–95

U.S. Postal Service
postal rates, 138–40
ZIP Codes, 208–10
U.S. Standard Time, 182
Universal Product Codes (UPC), 7–11
Universal Time, 182
Coordinated (UTC), 182
Universe, 196–98
UPC (Universal product codes), 7–11
Uranium-238 dating, 3–4
UTC (Coordinated Universal Time), 182
U-value, 100

Vision, 198
Vitamins and minerals, 116, 199–200
calcium, 17–19
Volt, 60–61, 116
Volume, xiv, 201–3
British measures, 202
Colonial measures, 201
metric equivalents for, 118–22
See also names of individual units of measure.

Watt, 60, 64, 116
Watt, James, 63–64
Wattage:
  appliances, 62
  light bulbs, 112–14
  microwave ovens, 123
Waves (electromagnetic), 150–51
  sound, 170–72
Weather. *See* Barometric pressure;
    Humidity; Rain; Snow; Wind.
Week, 20, 203
Weight, 203–6
  of common woods, 77–78
  Egyptian measures, 204
  metric measures, 117–22
  of ships, 159–60
Wilkinson, David, 158
Wind, 206–7
  Beaufort's wind scale, 207
  wind barometer table, 13–14
  wind chill factor, 37–38
Wine, 5, 116
  pH of, 134
Wood:
  firewood, 74–78
  lumber, 114–16

plywood, 136–37
weights of common woods, 77–
  78
WWV time signal, 182–3

X rays, 150–51

Yard (measure), 67–68, 110–11
  cubic, 121
  metric equivalent for, 117, 119,
    121–22
  square, 122
Yardage conversion chart, 68
Year, 20–21
  leap, 21
  mean solar, 181, 183
  meteorological, 152

ZIP Codes, 208–10
  used for determining R-value
    needs, 101–3
  used for determining Social Secu-
    rity numbers, 164
  zones, for, 209
"Z" time ("Zebra," "Zero," or
  "Zulu"), 89